DISCARDED FROM STOCK

21 0 01103447

NORTH STAFFORDSHIRE
POLYTECHNIC
LIBRARY

Fundamentals of Finite Elements for Engineers

Fundamentals of Finite Elements for Engineers

by

B. NATH

Lecturer in Civil Engineering,
Queen Mary College,
University of London

THE ATHLONE PRESS *of the University of London* 1974

Published by
THE ATHLONE PRESS
UNIVERSITY OF LONDON
at 4 Gower Street London WCI

Distributed by Tiptree Book Services Ltd
Tiptree, Essex

USA and Canada
Humanities Press Inc
New York

ISBN 0 485 11148 9

Set in 10/12pt Monophoto Times
at The Universities Press, Belfast,
and printed by photolithography
in Great Britain at the Pitman Press, Bath

To my Parents

Preface

Tremendous progress has been made in recent years in the finite element method of analysis. This essentially simple, elegant and powerful method has opened up a completely new vista to scientists and engineers who are now able to use it to analyse accurately, and with comparative ease, problems that were thought of as complex and intractable not so long ago. The method is still in process of evolution and, as an innovation in the context of scientific and engineering analysis, its full potential is yet to be realised.

Unfortunately, and in some ways perhaps inevitably, the finite element method has so far been confined mainly to postgraduate research and teaching. This, together with the advanced mathematical skill demanded by most texts available at present, is responsible for creating a curious *mystique* about the subject—a mystique which has the aura of a certain elitist exclusiveness that inhibits those with a modest mathematical repertoire from understanding what is after all a simple numerical method of analysis.

In this book I have tried to present the fundamentals of the method, in the simplest possible terms, to those familiar with the rudiments of matrix Algebra and computer programming.

The philosophy of the method and its *modus operandi* have been outlined in Chapter 1. In Chapters 2, 3 and 4 I have discussed the definitions, problem formulation and boundary conditions, respectively, by focussing attention onto simple beam structures with which most engineering and some science undergraduates are familiar. Experience shows that in general beginners find the assembly procedure of the 'overall stiffness matrix' most difficult to understand. I have therefore discussed this process in some detail in Chapter 3. Another notable hurdle that a beginner is likely to meet is the concept of 'rigid body' displacement and the related topic of 'boundary conditions'. To explain these in simple terms I have considered the trivial example of a bar 'structure' in Chapter 4, which I hope withstands rigorous scrutiny. The solution of some simple beam problems has been included in Chapter 5 to illustrate the method developed in Chapters 2–4.

I have devoted Chapters 6 and 7 to the solution of rigid and pin jointed frames, respectively. Appropriate transformation matrices and their role in the analysis of such structures has been discussed and, in each case, the method of solution illustrated by means of a detailed worked out example.

It is my belief that a part II Engineering or Science student at a University,

or one at an equivalent level elsewhere, ought to be able to assimilate the contents of Chapters 1–7 without difficulty.

Chapter 8 contains an introduction to the theory of linear elasticity. The general relationships from which the stiffness and other properties of finite elements are calculated, have been derived in Chapter 9. In this chapter I have discussed, with a deliberate emphasis on simplicity, the convergence and other criteria for choosing the assumed 'displacement function' of the element. Also, I have pointed out that a given finite element is characterised by what I have called its 'primary' matrices $[c]$, $[M]$ and $[N]$, since all the properties of the element can be found by manipulating these matrices.

The finite element solution of plates and shells by using flat elements has been discussed in Chapter 10. Here I have emphasised the point that although the primary matrices of these elements may appear formidable in size to the beginner, their automatic processing, however, does not present significant additional difficulty when compared with that of much simpler elements. I feel therefore that undergraduates in their final year of study should be encouraged to use these elements for solving simple flat plate problems. To this end I have included in this chapter a detailed solved example of this type, and a program for its solution in Chapter 15.

Chapter 11 deals with the finite element solution of the plane problems of elasticity. The detailed worked out example given in this chapter was specially chosen for the discretization problems it poses, as in many problems of this type that occur in practice, and also for the different kinds of conditions prescribed at the boundaries. Chapter 12 contains a brief discussion on the solution of three-dimensional elastic problems by using simple three-dimensional elements.

Solution of elasto-dynamic problems constitutes an important component of engineering analysis. The method of solution of these problems has been given in section 13.1. In various technical and scientific disciplines the solution of 'field' equations, discussed in section 13.2, is of considerable importance. I have used the simple method of 'elastic analogy' to solve a selection of these equations by finite elements, the main advantage of this approach being that the solution scheme is now identical to that for an analogous structural or elastic problem. A typical example of the general variational approach to these problems is given in Appendix 2. Chapter 14 will show, I hope, the simple and elegant manner in which the critical loads of structures can be calculated by using finite elements.

Chapter 15 is mainly on computer programming for finite elements. Three specific programs have been given, each accompanied by an explanatory note. In writing these programs I have aimed for clarity and simplicity, sometimes at the expense of brevity. Also, I have deliberately not included any program for beam-type problems, since, having devoted Chapters 2–7 to these relatively simple structures, I feel that the reader ought to be able to write suitable

programs for these by using the data scheme of section 15.4 and the *assembly block* of program 1.

Chapters 10–14 are intended mainly for postgraduate readers although undergraduates may also derive substantial benefit from them.

I am in the eternal debt of my brother Mr P. Nath, who collaborated with me in writing Chapters 9–11, for his numerous valuable suggestions and also for his encouragement.

I have derived considerable pleasure from writing this book. I must confess, however, that in the process I have come to appreciate, more than ever before, the difference between possessing a certain knowledge and being able to impart it simply and succintly.

Queen Mary College,
University of London

B. N.

Contents

Appendices

1 Introduction

1.1 General

Scientists and Engineers are often faced with practical physical problems whose solution by conventional analytical methods is either too difficult or even impossible. Consider, for instance, a three-dimensional elastic body which is acted upon by a set of externally applied forces. In order to evaluate the 'exact' response of the body to these forces, we must seek a 'closed form' solution of the equations which govern its deformation. However, due mainly to the complex geometrical configurations that practical problems usually have, it is exceedingly difficult and often impossible to obtain such a solution. Problems of this type arise all too frequently in various Scientific and Engineering disciplines. Faced with this, the natural recourse of the analyst is to seek a so-called 'numerical' solution to the problem. There is a number of numerical procedures available with which to attack an otherwise insoluble problem. The 'Finite Element' method is one such method. It is a relatively new method and it has a number of distinctive features which make it superior to most other methods.

1.2 The finite element method

In this method the body* is imagined to be actually broken up into a number of 'elements' of 'finite' dimensions, hence its name. If the body has n (= 1, 2, 3) dimensions of space, it is subdivided into a system of n-dimensional finite elements.

One-dimensional bodies will be subdivided into finite elements by means of 'nodes', as in Fig. 1.1a; lines and planes will be used for the subdivision of two- and three-dimensional bodies, as shown in Figs. 1.1b and c respectively. In one-dimensional bodies the resulting finite elements may have unequal lengths, while in two and three dimensions they may have unequal sizes as well as unlike shapes. In all cases, however, the finite elements representing the body will be 'interconnected' by means of 'nodes' as shown in Figs. 1.1a, b and c. Thus, in the finite element method of analysis the body will be replaced by a system of finite elements and the nodes connecting them.

The precise manner in which finite elements are attached to the nodes is best understood by referring to Fig. 1.2; here we have two plane finite elements of

* The general term 'body' will be used to mean a structure, a continuum or the domain of the problem.

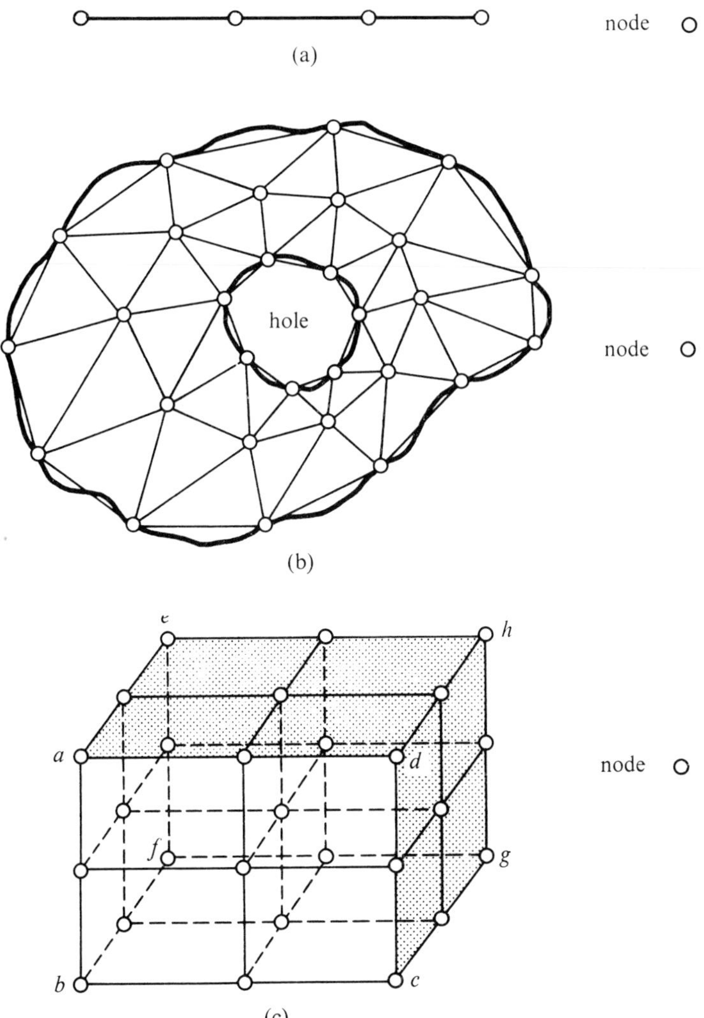

Fig. 1.1 (a) A one-dimensional body subdivided into 3 linear Finite elements. (b) A two-dimensional body with a hole subdivided into a system of plane triangular elements. (c) The three-dimensional body, a b c d e f g h, subdivided into 8 identical rectangular prism elements.

unit uniform thickness, one triangular and the other rectangular. Clearly, in Fig. 1.2a, they are separate and not attached to each other in any way. We shall think of the nodes as 'nut-and-bolt' devices which secure adjacent finite elements through their 'extremities'† and hold them together, as in Fig. 1.2b, such that the elements will come apart when the nodes are removed. Clearly

† The 'extremities' of a finite element are defined as its ends or corners through which the element is attached to the nodes, as in Fig. 1.2.

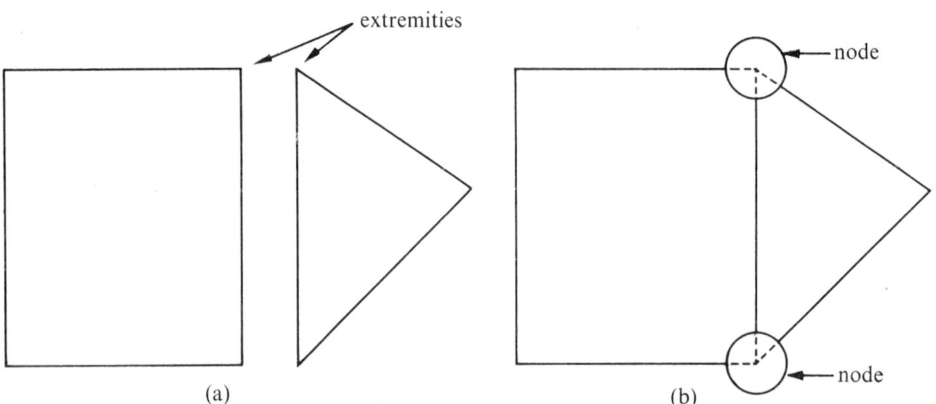

Fig. 1.2 (a) Two separate plane finite elements of unit uniform thickness. (b) The finite elements of (a) held together by means of nodes.

therefore, since they will come apart when the nodes securing them are removed, there is no physical continuity between adjacent finite elements.

The next step in this method of analysis is to determine the 'element stiffness matrix' of the individual elements representing the body. These will then be assembled to form the 'overall stiffness matrix' for the entire 'discretized' (i.e., broken-up) body by requiring that the continuity of displacements and equilibrium of forces prevail at all nodes in the finite element model of the body. This will lead to the matrix equation

$$[K]\{\delta\} = \{P\} \tag{1.1}$$

in which $[K]$ denotes the 'overall stiffness matrix' of the body. The 'overall force vector' $\{P\}$ lists the externally applied forces at all the nodes, while $\{\delta\}$ lists the displacements of all the nodes. Throughout this book [ ] and { } will denote square (or rectangular) matrices and vectors respectively.

An inspection of Eq. (1.1) shows that, qualitatively, $[K]$ represents the force required to produce unit displacement of the discretized body. Therefore, if we think of the finite element model of the body as an equivalent 'spring', then $[K]$ will obviously be a 'spring constant' representing its 'stiffness'. Thus, the finite element method is essentially one in which the analysis of the body is carried out from the point of view of its 'stiffness'. The concept of stiffness will be discussed in section 2.2.

For a given set of prescribed boundary conditions and external forces acting on the body, Eq. (1.1) can be solved uniquely for the nodal displacements, $\{\delta\}$, from which the stresses and strains within the body can subsequently be computed.

To summarise, the finite element solution of a given problem will require the

execution of the following operations in this order:

(1) 'Discretization' (subdivision) of the body into a system of finite elements.

(2) Derivation of the 'element stiffness matrix' and other properties for each individual element representing the body.

(3) Assembly of the 'overall stiffness matrix' $[K]$ and the 'overall force vector' $\{P\}$.

(4) Solution of Eq. (1.1) with prescribed boundary conditions to determine $\{\delta\}$, and

(5) Calculation of stresses and strains within the elements from the computed nodal displacements, $\{\delta\}$.

Practical scientific and engineering problems usually give rise to large $[K]$ matrices so that the use of a computer to solve Eq. (1.1) becomes inevitable. Simple programs can be written to 'mechanize' the above executions. Indeed, the finite element method, in conjunction with automatic computation, constitutes a very effective and elegant device for solving accurately complex physical problems, whose solution by any other means is either too difficult or even impossible.

Choice of element shape

Figs. 1.3 and 1.4 show some typical element shapes. When determining how to subdivide a given body, we are guided to a large extent by its geometry, particularly the shape of its boundaries both external and internal (the perimeter of the hole in Fig. 1.1b is a typical 'internal' boundary). In two-dimensional bodies it will be found that owing to their shapes, the triangular and quadrilateral elements are better able to deal with curved or awkward boundaries than do rectangular elements. In three dimensions the tetrahedron also has this advantage over the rectangular prism (Fig. 1.4). Depending on the shape of the body, sometimes it is more convenient in practice to adopt a 'mixed' subdivision in which more than one element shape is used. Fig. 1.5 shows such a combination representing a plane body.

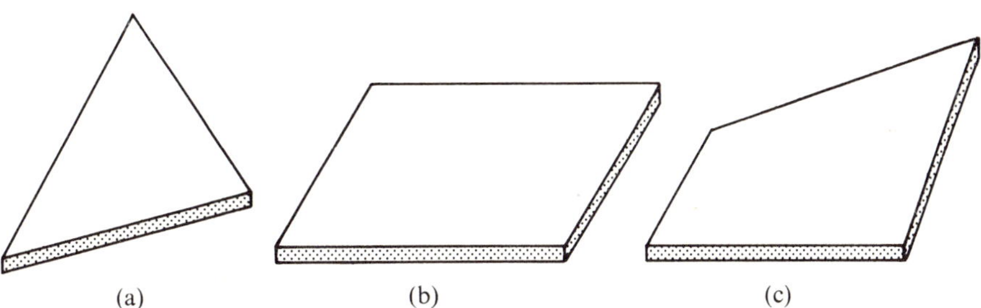

Fig. 1.3 Two-dimensional (plane) elements: (a) triangular, (b) rectangular and (c) quadrilateral.

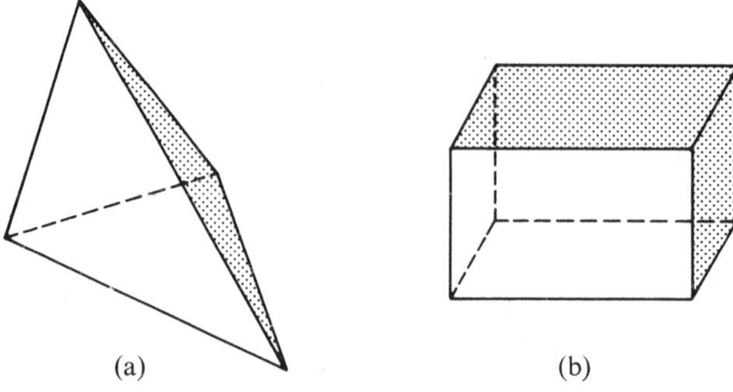

Fig. 1.4 Three-dimensional elements; (a) the tetrahedron and (b) the rectangular prism.

1.3 Why finite elements?

Why should anyone use the Finite Element method in preference to other numerical methods, in particular to the Finite Difference method which after all is well founded, older and reliable? The main points in favour of the finite element method over other methods are these:

(1) Owing to the flexibility of their sizes and shapes, finite elements are able to represent a given body, however complex its shape may be, more faithfully.

(2) Multiply-connected domains (i.e., bodies with one or more holes in them) or those with corners (Fig. 1.6) can be dealt with without difficulty.

(3) Problems involving variable material properties and/or variable geometry, do not present any additional difficulty. Geometrical and material non-linearities, even hereditary (i.e., time-dependent) material properties can be dealt with relatively easily.

(4) Problems of cause-effect relationships are formulated in terms of generalized 'forces' and 'displacements' which are related through the overall

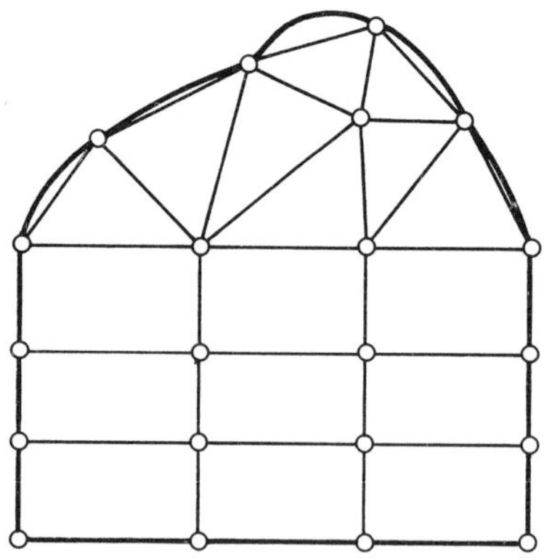

Fig. 1.5 A plane body subdivided into rectangular and triangular elements.

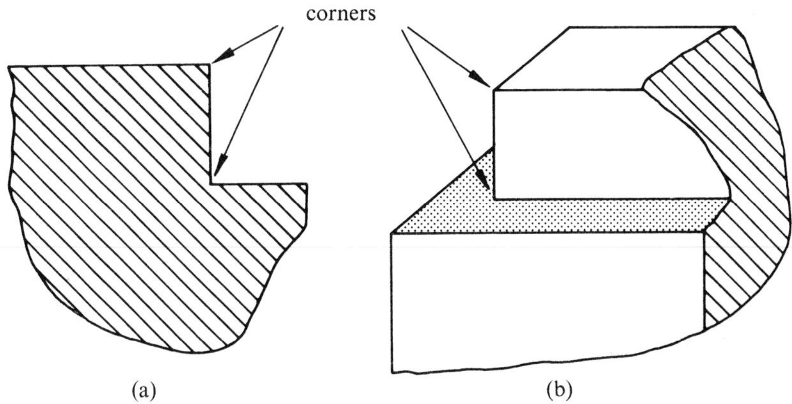

Fig. 1.6 Corners in (a) two dimensions and (b) in three dimensions.

stiffness matrix. This aspect of the finite element method facilitates and indeed simplifies the understanding of the problem and its solution.

(5) Boundary conditions are easily dealt with.

(6) The versatility and flexibility of the finite element method can be used very effectively to evaluate the cause-effect relationships in complex structural, continuum, field and other problems. Resulting accuracy is well comparable and often better than those arrived at via other analytical (if possible) or experimental methods.

Finite Difference treatment of 1, 3, 5 and particularly 2 is usually beset with considerable difficulties.

2 The concept of stiffness: the beam element

2.1 Assumptions

Unless otherwise stated, the following assumptions will remain valid throughout this book:

(1) The displacements within the deformed body are linearly related to the forces so that Hooke's law is obeyed.

(2) The displacements are small and are linearly related to the strains within the deformed body, and

(3) the geometrical and material properties of a finite element remain constant throughout it.

Bodies in which assumption (1) is violated are said to be 'materially non-linear'; such behaviour is exhibited by non-linearly elastic and plastic or viscoelastic materials.

Bodies complying with assumption (2) are said to be 'geometrically linear'. If, on the other hand, the displacements are sufficiently large to cause significant alteration of the geometry of the deformed body as compared to its undeformed state, then the body will behave in a 'geometrically non-linear' fashion. Geometrical non-linearity will be discussed in Chapter 14.

Assumption (3) implies that when the material and/or geometrical properties of the body vary from point to point, the average values of these properties over each element will be assumed to remain constant throughout that element. Thus, individual elements representing such a body will have different but constant values of these properties. In general, this assumption will be discarded if the variation of these properties can be represented by means of continuous functions (see section 9.8b).

2.2 The spring constant

Readers may have come across the term 'spring constant', often used in various technical disciplines. Its concept is fundamental in the finite element method of analysis, which is basically a 'stiffness' method as pointed out in Chapter 1.

Let u be the extension of the elastic, prismatic uniform beam of Fig. 2.1 when subjected to the axial force F. By definition, the 'spring constant' is the force required to produce unit displacement. In this particular case we can therefore show that if a and E denote the beam's area of cross section and Young's modulus respectively, then

$$\text{spring constant} = F/u = aE/h \tag{2.1}$$

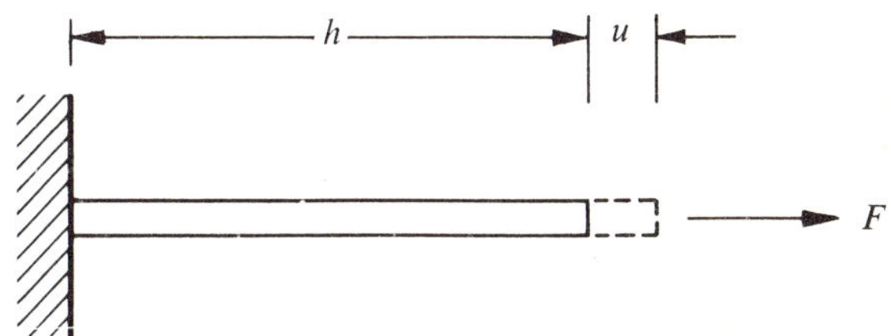

Fig. 2.1 A uniform prismatic beam, subjected to an axial force.

By virtue of assumption (1) (section 2.1), the quantity aE/h will always remain constant for a given beam. We could also apply, separately, other kinds of forces such as moments, shear etc. at the free end of the beam, each force producing its particular displacement. In each case, we could then compute the 'spring constant' as the corresponding force/displacement ratio.

For the sake of consistency, we shall now replace the term 'spring constant' by 'stiffness', so that the spring constant of Eq. (2.1) now becomes the 'axial stiffness' of the beam. Similarly, we have 'bending stiffness', 'torsional stiffness' etc.

2.3 Degrees of freedom

Consider the beam of Fig. 2.2a, a finite element subdivision of which is shown in Fig. 2.2b. Here, as in the rest of this book, the digits enclosed within circles are the 'node numbers', while those enclosed within squares are the 'element numbers'. In addition, the extremities of each element are identified by the letters a and b as shown (when it is convenient to do so, the extremities will be identified by digits rather than by letters).

Let us now subject this beam to a set of external forces (not shown in the diagram) which give rise to three kinds of internal (reactive) forces, namely, the axial force F, the transverse force Q and the bending moment M. Also, let u, v and θ be the respective displacements. Then, by isolating an element, say element number 2, we can represent the internal forces at the extremities of this element as shown in Fig. 2.2c.

[*Notation*. The first subscript in Fig. 2.2c identifies the extremity and the second the element to which the extremity belongs. Thus, F_{a2}, for example, is the internal axial force at extremity a of element number 2. Similarly, θ_{bi} is the rotation of extremity b of element number i, etc.]

Clearly, in the elements of Fig. 2.2b three kinds of forces and their respective displacements occur at each extremity. We shall therefore say that each element* of this beam has 3 'degrees of freedom'. Thus, the 'degrees of freedom' of a

* Strictly speaking, degrees of freedom should refer to the extremities (or nodes). Therefore, when we say that a finite element has m degrees of freedom, we actually mean that each of its extremities (or nodes) has m degrees of freedom.

finite element will be defined as the number of kinds of forces, and hence the number of kinds of displacements, occurring at each of its extremities.

2.4 The extremity and nodal vectors

The extremity vectors of a finite element refer to its extremities.

If a finite element, whose number is i, has the extremities $a, b, c, \ldots$, then its 'extremity force vectors' will be denoted by $\{P_{ai}\}, \{P_{bi}\}, \{P_{ci}\} \ldots$; $\{P_{ai}\}$, for example, lists all the *internal* forces at extremity a of element i; similarly $\{P_{bi}\}$ for extremity b of i, etc. In the particular case of Fig. 2.2c, the element has the extremities a and b, and $i = 2$. Consequently, its 'extremity force vectors' are

$$\{P_{a2}\} = \begin{Bmatrix} F_{a2} \\ Q_{a2} \\ M_{a2} \end{Bmatrix} = \{F_{a2} \quad Q_{a2} \quad M_{a2}\}^{\mathrm{T}} \tag{2.2a}$$

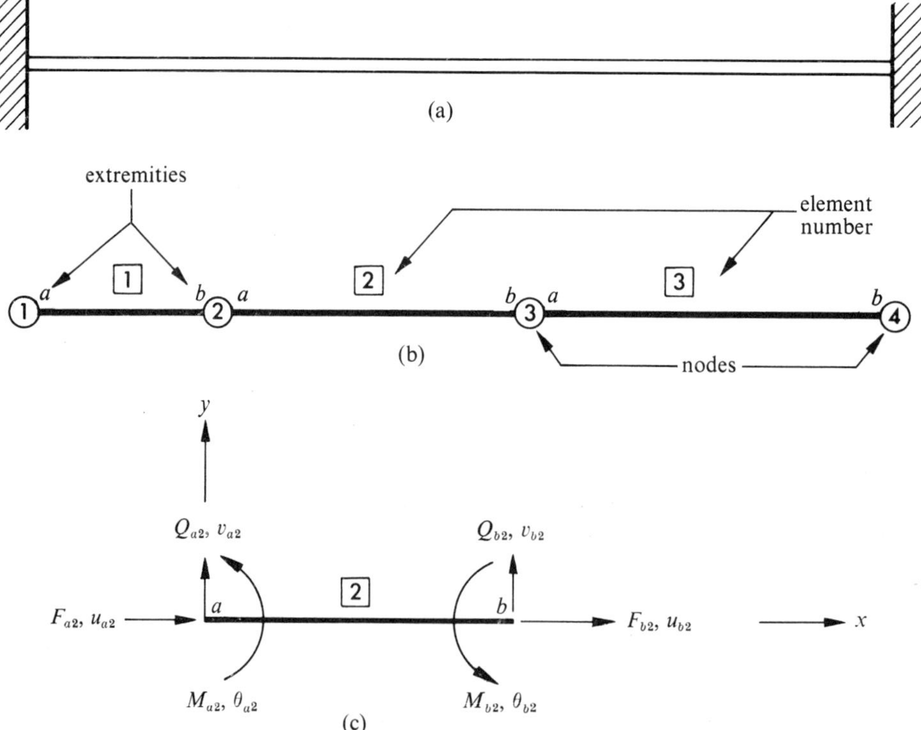

Fig. 2.2 (a) A beam, fixed at both ends. (b) A finite element model of (a). (c) Forces and displacements at the extremities of element number 2 (arrows indicate positive directions).

and $$\{P_{b2}\} = \{F_{b2} \quad Q_{b2} \quad M_{b2}\}^{\mathrm{T}} \tag{2.2b}$$

[*Notation*. The symbol $\{\ \}^{\mathrm{T}}$ denotes a 'transposed' vector (rows for columns). To conserve space, we shall often write lengthy vectors in their transposed form in this way.]

The 'extremity displacement vectors' of finite element i, whose extremities are a, b, c . . . , will be denoted by $\{\delta_{ai}\}$, $\{\delta_{bi}\}$, $\{\delta_{ci}\}$ . . . ; $\{\delta_{ai}\}$, for example, lists all the displacements occurring at extremity a of element i; similarly $\{\delta_{bi}\}$ for extremity b of i, and so on. In the particular case of Fig. 2.2c, we can therefore write

$$\{\delta_{a2}\} = \{u_{a2} \quad v_{a2} \quad \theta_{a2}\}^{\mathrm{T}} \tag{2.3a}$$

and $$\{\delta_{b2}\} = \{u_{b2} \quad v_{b2} \quad \theta_{b2}\}^{\mathrm{T}} \tag{2.3b}$$

The nodal vectors, on the other hand, refer to the nodes of the discretized body. Thus, the 'nodal force vector' $\{P_k\}$ lists all the *external* forces applied at node k, while the 'nodal displacement vector' $\{\delta_k\}$ lists all the displacements occurring at node k. In the particular case of Fig. 2.2b, where $k = 1, 2, 3, 4$, imagine that we have applied at node k the axial force F_k, the transverse force Q_k and the bending moment M_k. Then, by definition

$$\{P_k\} = \{F_k \quad Q_k \quad M_k\}^{\mathrm{T}} \tag{2.4a}$$

Also, if u_k, v_k and θ_k are the axial, transverse and rotational displacements at node k, respectively, then by definition

$$\{\delta_k\} = \{u_k \quad v_k \quad \theta_k\}^{\mathrm{T}} \tag{2.4b}$$

It is important to observe that the listing sequences of all extremity and nodal vectors must be identical. Eqs. (2.2)–(2.4) show that in the sequence adopted for the beam of Fig. 2.2a, the axial, transverse shear and bending modes are listed in this sequence. Having adopted this sequence, it would have been incorrect to write $\{\delta_k\}$, for example, as

$$\{\delta_k\} = \{\theta_k \quad u_k \quad v_k\}^{\mathrm{T}}$$

since the sequence here differs from that adopted for the beam.

Also, observe that all extremity forces are *internal* forces, while all nodal forces are *externally applied* forces.

2.5 The element stiffness matrix

According to assumption (1) (section 2.1), the forces within the body are linearly related to the displacements. A very important consequence of this linearity is the well known 'principle of superposition'. Using this, the total 'effect' due to the simultaneous action of several 'causes' can be expressed as the sum of effects due to each cause acting separately. Now, consider a uniform beam

element whose number is i and whose degrees of freedom are the same as those shown in Fig. 2.2c. Then, treating the extremity displacements and forces as causes and effects respectively, and using this principle, we can write the following linear force-displacement relationships for the beam element i:

$$Q_{ai} = A_{11}v_{ai} + A_{12}\theta_{ai} + A_{13}v_{bi} + A_{14}\theta_{bi} \quad (2.5a)$$

$$M_{ai} = A_{21}v_{ai} + A_{22}\theta_{ai} + A_{23}v_{bi} + A_{24}\theta_{bi} \quad (2.5b)$$

$$Q_{bi} = A_{31}v_{ai} + A_{32}\theta_{ai} + A_{33}v_{bi} + A_{34}\theta_{bi} \quad (2.5c)$$

and

$$M_{bi} = A_{41}v_{ai} + A_{42}\theta_{ai} + A_{43}v_{bi} + A_{44}\theta_{bi} \quad (2.5d)$$

in which the constants A_{11}, A_{12} etc. are the 'stiffness coefficients' of element i. We have deliberately excluded the axial forces and displacements from Eqs. (2.5), since, by virtue of assumed geometrical linearity (section 2.1), the transverse stiffnesses of Eqs. (2.5) are not affected by the axial forces. The axial and torsional stiffnesses will be derived independently, as we shall soon see.

Writing the above equations in the matrix form, we have

$$\begin{Bmatrix} Q_{ai} \\ M_{ai} \\ Q_{bi} \\ M_{bi} \end{Bmatrix} = \begin{bmatrix} A_{11} & A_{12} & A_{13} & A_{14} \\ A_{21} & A_{22} & A_{23} & A_{24} \\ A_{31} & A_{32} & A_{33} & A_{34} \\ A_{41} & A_{42} & A_{43} & A_{44} \end{bmatrix} \begin{Bmatrix} v_{ai} \\ \theta_{ai} \\ v_{bi} \\ \theta_{bi} \end{Bmatrix} \quad (2.6a)$$

The square matrix of Eq. (2.6a), which is made up of the stiffness coefficients of element i, is called the 'element stiffness matrix' of the beam element i. Denoting this matrix by $[K_i]$, Eq. (2.6a) can be written in the symbolic form as

$$\{\bar{P}\} = [K_i]\{\bar{\delta}\} \quad (2.6b)$$

in which $\{\bar{P}\}$ lists all the extremity forces of the element, while $\{\bar{\delta}\}$ lists all its extremity displacements (see section 2.7).

We can now generalize and define the 'element stiffness matrix' $[K_i]$ of any finite element i as a matrix which relates the element's extremity forces to its extremity displacements in accordance with Eq. (2.6b). $[K_i]$ is always a symmetric matrix owing to the reciprocity of causes and effects in linear systems. It is easily verified that if i has n extremities and m degrees of freedom, then its $[K_i]$ will have $n \times m$ columns and $n \times m$ rows.

Writing Eq. (2.1) alternatively as

$$F = (\text{axial stiffness})u$$

and comparing this with Eq. (2.6b), it is clear that both these equations have the same physical interpretation in that they both define the concept of stiffness; however, whereas the above equation is concerned with a single force, Eq. (2.6b) represents in the matrix form the stiffness related to various forces.

2.6 Derivation of the stiffness coefficients

The stiffness coefficients A_{11}, A_{12} etc. for the beam element can be derived explicitly from the elementary considerations of structural mechanics (a general method for deriving the stiffness and other properties of finite elements will be given in Chapter 9).

If we set $v_{ai} = 1$ and all other displacements equal to zero in Eq. (2.5a), for example, then, by the definition of section 2.2 the stiffness coefficient A_{11} will be given by

$$A_{11} = Q_{ai}$$

In other words, the stiffness coefficient A_{11} is equal to the shear force that obtains at extremity a, when element i is forced to comply with the following displacements

$$v_{ai} = 1 \quad \text{and} \quad v_{bi} = \theta_{ai} = \theta_{bi} = 0.$$

Fig. 2.3a shows the attitude of the beam element when it is forced to comply

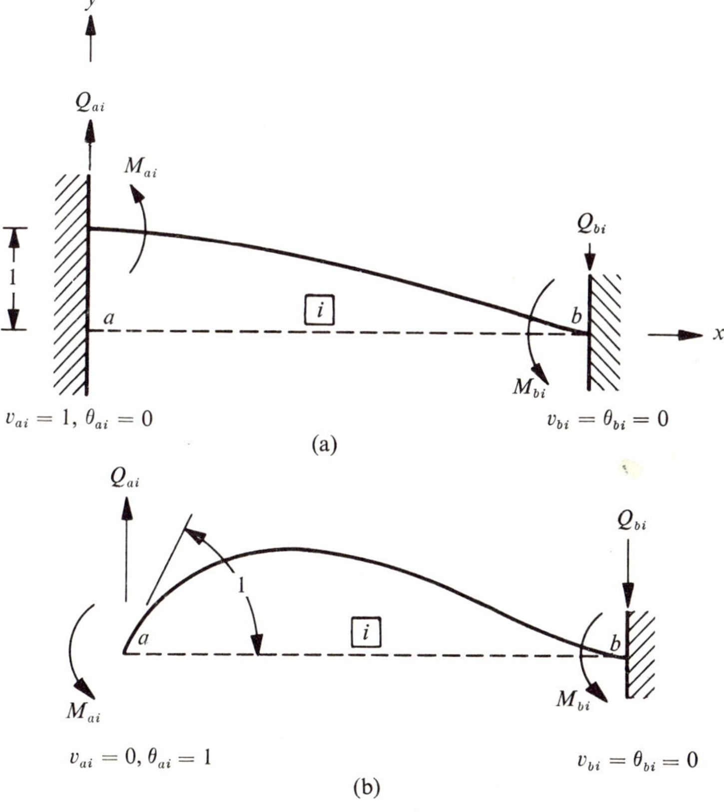

Fig. 2.3

with the above displacements. It is a simple matter now to show from the slope-deflection method, for example, that in this attitude of the element

$$A_{11} = Q_{ai} = 12E_iI_i/h_i^3$$

in which E_i, I_i and h_i are Young's modulus, second moment of area and length of the element, respectively.

Similarly, when we set

$$v_{ai} = v_{bi} = \theta_{bi} = 0 \quad \text{and} \quad \theta_{ai} = 1$$

the element takes up the attitude shown in Fig. 2.3b; Eq. (2.5a) now gives

$$A_{12} = Q_{ai}$$

indicating that A_{12} is equal to the shear force that obtains at extremity a, when element i is constrained to the above displacements. The slope-deflection method now gives

$$A_{12} = 6E_iI_i/h_i^2$$

We can obviously continue this process to calculate the explicit expressions for all the coefficients of Eq. (2.6a).

Axial stiffness

Owing to the assumed geometrical linearity mentioned before, these stiffnesses do not affect displacements other than their own and therefore can be derived independently. Let s_{11}, s_{12}, s_{21} and s_{22} be the axial stiffness coefficients of element i. Then, using the principle of superposition, as in Eq. (2.5), we can write

$$F_{ai} = s_{11}u_{ai} + s_{12}u_{bi} \tag{2.7a}$$

and

$$F_{bi} = s_{21}u_{ai} + s_{22}u_{bi} \tag{2.7b}$$

With $u_{bi} = 1$ and $u_{ai} = 0$ (Fig. 2.4), for instance, we have from Eq. (2.7a)

$$s_{12} = F_{ai}$$

Then, using Eq. (2.1) and noting that the arrows in Fig. 2.2c indicate positive directions, we can demonstrate that

$$s_{11} = s_{22} = -s_{12} = -s_{21} = a_iE_i/h_i$$

in which a_i is the area of cross section of the beam.

Torsional stiffness

Let us now introduce into Fig. 2.2c a fourth degree of freedom in the form of torsional forces and displacements as shown in Figs. 2.5 and 2.6. The torsional forces T_{ai} and T_{bi} and their respective displacements (angles of twist) ϕ_{ai} and ϕ_{bi}

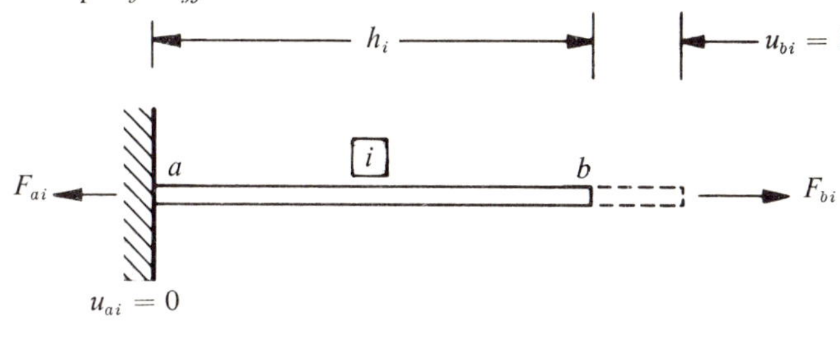

Fig. 2.4

lie on planes normal to the axis of the element. The torsional forces, like the axial forces, do not affect displacements other than their own (by virtue of assumption 2, section 2.1). As before, we can therefore write

$$T_{ai} = t_{11}\phi_{ai} + t_{12}\phi_{bi}$$

and

$$T_{bi} = t_{21}\phi_{ai} + t_{22}\phi_{bi}$$

in which t_{11}, t_{12} etc. are the 'torsional stiffness coefficients'. Setting $\phi_{ai} = 0$ and $\phi_{bi} = 1$, as in Fig. 2.5, we have from the first of the above equations

$$t_{12} = T_{ai}$$

indicating that t_{12} is equal to the torque obtained at extremity a when the element is constrained to the above displacements. The value of T_{ai}, consistent with these displacements, can now be found from the elementary theory of strength of materials. The reader may verify that

$$t_{11} = t_{22} = -t_{12} = -t_{21} = G_iJ_i/h_i$$

$$\begin{Bmatrix} T_{ai} \\ F_{ai} \\ Q_{ai} \\ M_{ai} \\ \hdashline T_{bi} \\ F_{bi} \\ Q_{bi} \\ M_{bi} \end{Bmatrix} = \frac{E_i}{h_i^3} \left[\begin{array}{cccc:cccc} \beta_i h_i^2 & 0 & 0 & 0 & -\beta_i h_i^2 & 0 & 0 & 0 \\ 0 & a_i h_i^2 & 0 & 0 & 0 & -a_i h_i^2 & 0 & 0 \\ 0 & 0 & 12I_i & 6I_i h_i & 0 & 0 & -12I_i & 6I_i h_i \\ 0 & 0 & 6I_i h_i & 4I_i h_i^2 & 0 & 0 & -6I_i h_i & 2I_i h_i^2 \\ \hdashline -\beta_i h_i^2 & 0 & 0 & 0 & \beta_i h_i^2 & 0 & 0 & 0 \\ 0 & -a_i h_i^2 & 0 & 0 & 0 & a_i h_i^2 & 0 & 0 \\ 0 & 0 & -12I_i & -6I_i h_i & 0 & 0 & 12I_i & -6I_i h_i \\ 0 & 0 & 6I_i h_i & 2I_i h_i^2 & 0 & 0 & -6I_i h_i & 4I_i h_i^2 \end{array}\right] \begin{Bmatrix} \phi_{ai} \\ u_{ai} \\ v_{ai} \\ \theta_{ai} \\ \hdashline \phi_{bi} \\ u_{bi} \\ v_{bi} \\ \theta_{bi} \end{Bmatrix} \quad (2.8)$$

$$\beta_i = G_iJ_i/E_i$$

in which G_i is the modulus of rigidity of i. The quantity J_i, which depends on

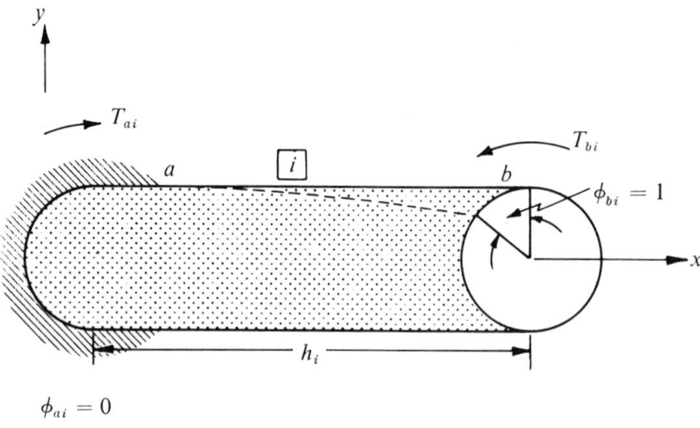

Fig. 2.5

the cross sectional shape of the element, is called the 'torsional stiffness factor' of i. For circular sections J_i is also called the 'polar second moment of area'.

Having thus calculated the various stiffnesses, we shall now put them in the matrix form of Eq. (2.8) (ignore the dotted lines for the time being). Since by definition (section 2.7) the left and right hand vectors of this equation are $\{\bar{P}\}$ and $\{\bar{\delta}\}$, respectively, it follows from Eq. (2.6b) that the square matrix of Eq. (2.8) is in fact the $[K_i]$ matrix of the uniform beam element i, which has 4 degrees of freedom as in Fig. 2.6.

In order to analyze a given problem by finite elements, the $[K_i]$'s of all elements into which the body is subdivided, will be calculated to begin with. These will then be assembled, by the methods to be discussed in the next chapter, to form the so called 'overall stiffness matrix' for the entire body.

2.7 Notation

The following notation and definitions will be valid for all finite elements to be used in this book:

The 'extremity' vectors

(1) The 'extremity force vector' $\{P_{ai}\}$ lists all the *internal* forces at extremity a of element i. If $a, b, c, \ldots$ are the extremities of element i, then its extremity force vectors will be $\{P_{ai}\}, \{P_{bi}\}, \{P_{ci}\} \ldots$.

(2) The 'extremity displacement vector' $\{\delta_{ai}\}$ lists all the displacements at extremity a of element i. If $a, b,$ c . . . are the extremities of element i, then its extremity displacement vectors will be $\{\delta_{ai}\}, \{\delta_{bi}\}, \{\delta_{ci}\} \ldots$.

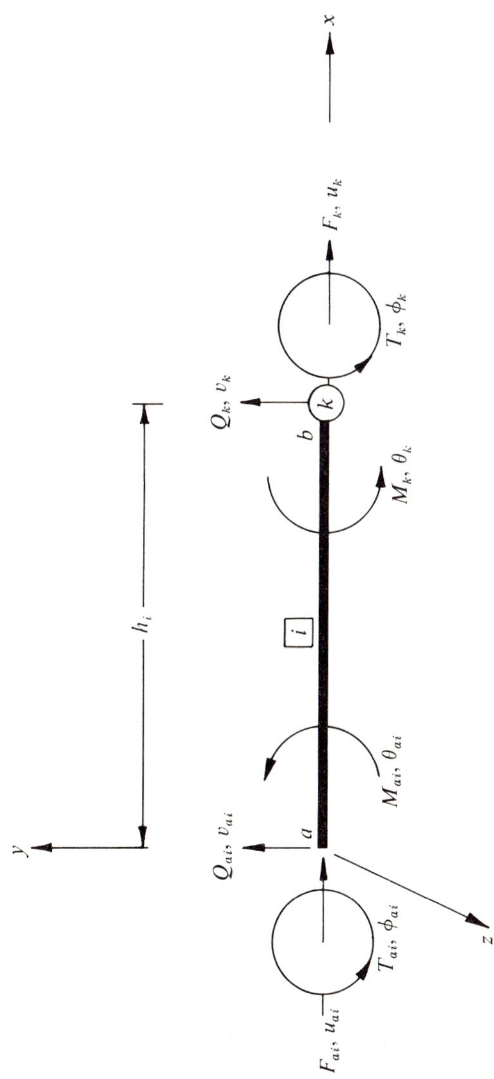

Fig. 2.6 Forces and displacements in a 4-degree-of-freedom beam element; arrows indicate positive directions.

The 'element' vectors

(1) The 'element force vector' $\{\bar{P}\}$ lists all the extremity forces of the element; thus,

$$\{\bar{P}\} = \begin{Bmatrix} \{P_{ai}\} \\ \{P_{bi}\} \\ \cdot \\ \cdot \\ \cdot \end{Bmatrix}$$

(2) The 'element displacement vector' $\{\bar{\delta}\}$ lists all the extremity displacements of the element; thus

$$\{\bar{\delta}\} = \begin{Bmatrix} \{\delta_{ai}\} \\ \{\delta_{bi}\} \\ \cdot \\ \cdot \\ \cdot \end{Bmatrix}$$

The 'nodal' vectors

(1) The 'nodal force vector' $\{P_k\}$ lists all the *external* forces applied at node k.
(2) The 'nodal displacement vector' $\{\delta_k\}$ lists all the displacements of node k.

The 'overall' vectors

(1) The 'overall force vector' $\{P\}$ lists all the nodal forces applied to the discretized body; thus, if the entire discretized body has N nodes, then

$$\{P\} = \begin{Bmatrix} \{P_1\} \\ \{P_2\} \\ \cdot \\ \cdot \\ \cdot \\ \{P_N\} \end{Bmatrix}$$

(2) The 'overall displacement vector' $\{\delta\}$ lists all the nodal displacements of the discretized body; thus

$$\{\delta\} = \begin{Bmatrix} \{\delta_1\} \\ \{\delta_2\} \\ \cdot \\ \cdot \\ \cdot \\ \{\delta_N\} \end{Bmatrix}$$

3 Assembly of the overall stiffness matrix

Having calculated the $[K_i]$'s of individual elements into which the body is subdivided, the next step is to assemble these to form what is called the 'overall stiffness matrix' for the entire discretized body. This is done by ensuring that the equilibrium and compatibility conditions are satisfied at all the nodes within the discretized body. The assembly procedure, as well as some typical examples will be given in this chapter.

3.1 The compatibility and equilibrium conditions

It is a fundamental requirement of elasto-mechanics that the displacements within the deformed body be 'compatible', i.e., continuous. Displacement compatibility and its implications will be discussed in detail from the elastic and finite element points of view in sections 8.8 and 9.5 respectively. In the meantime, for the purpose of assembling the overall stiffness matrix we shall merely require that continuity of displacements prevails only at the nodes.

As an example of nodal compatibility, consider the discretized beam of Fig. 2.2b. In order to be continuous, the displacements of extremity b of element 1 must be equal to those of extremity a of element 2; each of these displacements must, in turn, be equal to those of node 2 to which both these extremities are attached. That is

$$u_2 = u_{b1} = u_{a2}$$

$$v_2 = v_{b1} = v_{a2}$$

and

$$\theta_2 = \theta_{b1} = \theta_{a2}$$

or, writing these in the matrix notation of section 2.7,

$$\{\delta_2\} = \{\delta_{b1}\} = \{\delta_{a2}\} \tag{3.1a}$$

Similarly, for this beam

$$\{\delta_1\} = \{\delta_{a1}\} \tag{3.1b}$$

$$\{\delta_3\} = \{\delta_{a3}\} = \{\delta_{b2}\} \tag{3.1c}$$

and

$$\{\delta_4\} = \{\delta_{b3}\} \tag{3.1d}$$

The conditions expressed by these equations, when enforced, will ensure the continuity of displacements of the discretized beam.

For the loaded structure or continuum to be in equilibrium, the external forces applied at the nodes must be equal to their reactions. Following the notation of section 2.7, the external force that may be applied at node 2

(Fig. 2.2b), for example, is $\{P_2\}$. For equilibrium, $\{P_2\}$ must be equal to the reactions set up at the extremities attached to node 2; that is

$$\{P_2\} = \{P_{b1}\} + \{P_{a2}\}$$

Similarly, for this beam

$$\{P_1\} = \{P_{a1}\}$$

$$\{P_3\} = \{P_{a3}\} + \{P_{b2}\}$$

and

$$\{P_4\} = \{P_{b3}\}$$

We shall now put the above four equations in the following matrix form:

$$\begin{Bmatrix} \{P_1\} \\ \{P_2\} \\ \{P_3\} \\ \{P_4\} \end{Bmatrix} = \begin{bmatrix} 1 & 0 & 0 & 0 & 0 & 0 \\ 0 & 1 & 1 & 0 & 0 & 0 \\ 0 & 0 & 0 & 1 & 1 & 0 \\ 0 & 0 & 0 & 0 & 0 & 1 \end{bmatrix} \begin{Bmatrix} \{P_{a1}\} \\ \{P_{b1}\} \\ \{P_{a2}\} \\ \{P_{b2}\} \\ \{P_{a3}\} \\ \{P_{b3}\} \end{Bmatrix} \qquad (3.2)$$

We have assumed here that the external forces are applied only at the nodes. Consequently, when subdividing a given body into finite elements, a node must be provided at each external point load. This, however, need not be so when equivalent force vectors are used (see sections 5.2, 9.3 and 9.7).

3.2 The element submatrices

The assembly procedure, particularly its mechanization by using the computer, is considerably simplified by partitioning the $[K_i]$'s of finite elements representing the body into what we shall call their 'element submatrices'. This will be done on an extremity-by-extremity basis such that, if element i has n extremities then, its $[K_i]$ will be partitioned into a total of n^2 square submatrices of equal size. The element submatrices of i will be denoted by $S(i, J, K)$* in which J and K take on the values $J = 1, 2, \ldots n$ and $K = 1, 2, \ldots n$. It is easily verified that if the element's degree of freedom is m, then each of its $S(i, J, K)$'s will be a square matrix of size $m \times m$.

As an example,† consider the element submatrices of the beam element of Fig. 2.6, whose $[K_i]$ is defined by Eq. (2.8). Now, if we partition this matrix along the dotted lines as shown, then this obviously subdivides $[K_i]$ into four submatrices of equal size. Since this element has two extremities, J and K vary as $J = 1, 2$ and $K = 1, 2$; consequently, the element submatrices of this

* The notation $S(i, J, K)$ has been chosen in place of the conventional S_{iJK} since the former is consistent with the programming language of Chapter 15.

† See section 3.5 for another example.

element are

$$S(i, 1, 1), \quad S(i, 1, 2), \quad S(i, 2, 1) \text{ and } S(i, 2, 2)$$

and we can write $[K_i]$ in terms of these submatrices as

$$[K_i] = \begin{bmatrix} S(i, 1, 1) & S(i, 1, 2) \\ S(i, 2, 1) & S(i, 2, 2) \end{bmatrix} \tag{3.3}$$

Comparing Eq. (3.3) with the two-dimensional array of Eq. (2.8), it is clear that for the element of Fig. 2.6 we have by definition

$$S(i, 1, 1) = E_i \begin{bmatrix} \beta_i/h_i & 0 & 0 & 0 \\ 0 & a_i/h_i & 0 & 0 \\ 0 & 0 & 12I_i/h_i^3 & 6I_i/h_i^2 \\ 0 & 0 & 6I_i/h_i^2 & 4I_i/h_i \end{bmatrix} \tag{3.4a}$$

$$S(i, 1, 2) = E_i \begin{bmatrix} -\beta_i/h_i & 0 & 0 & 0 \\ 0 & -a_i/h_i & 0 & 0 \\ 0 & 0 & -12I_i/h_i^3 & 6I_i/h_i^2 \\ 0 & 0 & -6I_i/h_i^2 & 2I_i/h_i \end{bmatrix} \tag{3.4b}$$

and

$$S(i, 2, 2) = E_i \begin{bmatrix} \beta_i/h_i & 0 & 0 & 0 \\ 0 & a_i/h_i & 0 & 0 \\ 0 & 0 & 12I_i/h_i^3 & -6I_i/h_i^2 \\ 0 & 0 & -6I_i/h_i^2 & 4I_i/h_i \end{bmatrix} \tag{3.4c}$$

while, owing to the symmetry of $[K_i]$, $S(i, 1, 2)$ and $S(i, 2, 1)$ are transposes of each other. It also follows that Eq. (2.8) can be written in terms of the above $S(i, J, K)$'s as

$$\begin{Bmatrix} \{P_{ai}\} \\ \{P_{bi}\} \end{Bmatrix} = \begin{bmatrix} S(i, 1, 1) & S(i, 1, 2) \\ S(i, 2, 1) & S(i, 2, 2) \end{bmatrix} \begin{Bmatrix} \{\delta_{ai}\} \\ \{\delta_{bi}\} \end{Bmatrix} \tag{3.5}$$

3.3 Assembly of the overall stiffness matrix of a beam

Clearly, the $S(i, J, K)$'s of the elements of Fig. 2.2b will be obtained explicitly merely by deleting from Eqs. (3.4) the rows and columns which refer to torsion. Then, setting $i = 1$ in Eq. (3.5), the symbolic force-displacement relationship for element 1 (Fig. 2.2b) is found to be

$$\begin{Bmatrix} \{P_{a1}\} \\ \{P_{b1}\} \end{Bmatrix} = \begin{bmatrix} S(1, 1, 1) & S(1, 1, 2) \\ S(1, 2, 1) & S(1, 2, 2) \end{bmatrix} \begin{Bmatrix} \{\delta_{a1}\} \\ \{\delta_{b1}\} \end{Bmatrix}$$

or, using Eqs. (3.1a) and (3.1b), we obtain

$$\begin{Bmatrix} \{P_{a1}\} \\ \{P_{b1}\} \end{Bmatrix} = \begin{bmatrix} S(1,1,1) & S(1,1,2) \\ S(1,2,1) & S(1,2,2) \end{bmatrix} \begin{Bmatrix} \{\delta_1\} \\ \{\delta_2\} \end{Bmatrix} \tag{3.6a}$$

Similarly, for elements 2 and 3 we can show that

$$\begin{Bmatrix} \{P_{a2}\} \\ \{P_{b2}\} \end{Bmatrix} = \begin{bmatrix} S(2,1,1) & S(2,1,2) \\ S(2,2,1) & S(2,2,2) \end{bmatrix} \begin{Bmatrix} \{\delta_2\} \\ \{\delta_3\} \end{Bmatrix} \tag{3.6b}$$

and

$$\begin{Bmatrix} \{P_{a3}\} \\ \{P_{b3}\} \end{Bmatrix} = \begin{bmatrix} S(3,1,1) & S(3,1,2) \\ S(3,2,1) & S(3,2,2) \end{bmatrix} \begin{Bmatrix} \{\delta_3\} \\ \{\delta_4\} \end{Bmatrix} \tag{3.6c}$$

We shall now put the above equations in the following matrix form:

$$\begin{Bmatrix} \{P_{a1}\} \\ \{P_{b1}\} \\ \{P_{a2}\} \\ \{P_{b2}\} \\ \{P_{a3}\} \\ \{P_{b3}\} \end{Bmatrix} = \begin{bmatrix} S(1,1,1) & S(1,1,2) & 0 & 0 \\ S(1,2,1) & S(1,2,2) & 0 & 0 \\ 0 & S(2,1,1) & S(2,1,2) & 0 \\ 0 & S(2,2,1) & S(2,2,2) & 0 \\ 0 & 0 & S(3,1,1) & S(3,1,2) \\ 0 & 0 & S(3,2,1) & S(3,2,2) \end{bmatrix} \begin{Bmatrix} \{\delta_1\} \\ \{\delta_2\} \\ \{\delta_3\} \\ \{\delta_4\} \end{Bmatrix} \tag{3.7}$$

Then, from Eqs. (3.2) and (3.7) it immediately follows that

$$\begin{Bmatrix} \{P_1\} \\ \{P_2\} \\ \{P_3\} \\ \{P_4\} \end{Bmatrix} = \begin{bmatrix} 1 & 0 & 0 & 0 & 0 & 0 \\ 0 & 1 & 1 & 0 & 0 & 0 \\ 0 & 0 & 0 & 1 & 1 & 0 \\ 0 & 0 & 0 & 0 & 0 & 1 \end{bmatrix}$$

$$\times \begin{bmatrix} S(1,1,1) & S(1,1,2) & 0 & 0 \\ S(1,2,1) & S(1,2,2) & 0 & 0 \\ 0 & S(2,1,1) & S(2,1,2) & 0 \\ 0 & S(2,2,1) & S(2,2,2) & 0 \\ 0 & 0 & S(3,1,1) & S(3,1,2) \\ 0 & 0 & S(3,2,1) & S(3,2,2) \end{bmatrix} \begin{Bmatrix} \{\delta_1\} \\ \{\delta_2\} \\ \{\delta_3\} \\ \{\delta_4\} \end{Bmatrix} \tag{3.8}$$

From section 2.7 the left and right hand vectors of this equation are recognized as the overall force and displacement vectors, respectively, in this case of the beam of Fig. 2.2b. We can therefore write this equation in the symbolic form as

$$\{P\} = [K]\{\delta\} \tag{3.9}$$

in which we can show by performing the multiplication of Eq. (3.8) that

$$[K] = \begin{array}{c} \\ 1 \\ 2 \\ \\ 3 \\ \\ 4 \end{array}\begin{array}{c} \begin{array}{cccc} 1 & 2 & 3 & 4 \end{array} \\ \begin{bmatrix} S(1,1,1) & S(1,1,2) & 0 & 0 \\ S(1,2,1) & S(1,2,2) + S(2,1,1) & S(2,1,2) & 0 \\ 0 & S(2,2,1) & S(2,2,2) + S(3,1,1) & S(3,1,2) \\ 0 & 0 & S(3,2,1) & S(3,2,2) \end{bmatrix} \end{array} \tag{3.9a}$$

The $[K]$ of Eq. (3.9a) will be defined as the 'overall stiffness matrix' of the beam of Fig. 2.2a. Clearly, $[K]$ here is in terms of the $S(i, J, K)$'s, which can now be converted from Eqs. (3.4) to give $[K]$ in terms of the material and geometrical properties of the elements which subdivide the beam.

A comparison shows that whereas Eq. (2.6b) specifies the force-displacement relationship of element i in terms of its extremity forces and displacements, Eq. (3.9) states on the other hand, the force-displacement relationship of the entire discretized beam in terms of the nodal forces and displacements. To generalise, *in all finite element problems it is most important to observe this distinction between* $[K_i]$ *and* $[K]$.

It is a simple matter now to verify that if the discretized beam has a total of N nodes and 3 degrees of freedom (as in Fig. 2.2c), then

$$\{P\} = \{F_1\ Q_1\ M_1 \quad F_2\ Q_2\ M_2 \cdots F_N\ Q_N\ M_N\}^{\mathrm{T}} \tag{3.10}$$

and

$$\{\delta\} = \{u_1\ v_1\ \theta_1 \quad u_2\ v_2\ \theta_2 \cdots u_N\ v_N\ \theta_N\}^{\mathrm{T}} \tag{3.11}$$

When the external force vector $\{P\}$ is specified, Eq. (3.9) can be solved to give the nodal displacements $\{\delta\}$, from which the internal forces (such as moments, shear forces etc.) can be calculated. However, before attempting a solution to Eq. (3.9), we must modify $[K]$ by introducing into it the prescribed 'boundary conditions', since it is only then that this equation will have a unique solution. This aspect will be discussed in the next chapter.

To generalize, when dealing with the finite element solution of a given problem, our immediate objective would be to express the nodal force-displacement relationship of the entire discretized body in terms of an equation similar to Eq. (3.9), in which $[K]$ denotes the 'overall stiffness matrix' of the whole body. Then, having incorporated the prescribed boundary conditions into $[K]$ and having specified $\{P\}$, this equation will be solved uniquely for the nodal displacements $\{\delta\}$ from which the stresses and strains within the body can be calculated.

$[K]$ will always be found to be a square symmetric matrix, as we might have anticipated from the reciprocal theorem. Furthermore, if the discretized

body has a total of N nodes and if the degree of freedom is m, then it is easily verified that $[K]$ will have $N \times m$ rows and hence $N \times m$ columns.

Since the $[K_i]$ of each element is in terms of its individual geometrical and material properties (e.g., Eq. 2.8), it will be observed that in the finite element solution of problems in which these properties vary, the body will be subdivided into elements having different properties. Such problems therefore do not pose any additional difficulty. This is a very important feature of the finite element method of analysis.

3.4 A simple general method for assembling the overall stiffness matrix

Although instructive, the assembly procedure given in section 3.3 is unfortunately not suitable for dealing with complex problems, particularly when the body has a two or three dimensional configuration. This is because in such bodies the derivation of an equation similar to Eq. (3.2) becomes a tedious and often wasteful exercise. More often than not, the overall stiffness matrix is automatically assembled by the computer. It is important therefore that we have at our disposal a simple, efficient and general method that is easily programmed for the computer. Such a method will be given in this section; it will enable us to assemble the $[K]$ matrix of any discretized body merely by inspection, regardless of the complexity of the body or of the finite elements into which it is subdivided.

In this method we shall imagine that the extremities of every element of the discretized body 'visit' each other. To be more specific, if the body is subdivided into the elements $i = 1, 2, 3 \ldots$, then all the extremities of i will be imagined to 'visit' each other; the extremities of i will not, however, 'visit' any extremity that does not belong to i. Then, the $[K]$ matrix can be assembled in terms of the $S(i, J, K)$'s by following these simple rules:

(1) If extremity J is attached to node L and extremity K to node M and if both J and K belong to element i, then, when J visits K the submatrix $S(i, J, K)$ will appear in row L and column M of $[K]$.

(2) If extremity J of element i is attached to node L, then, when J 'visits itself' the submatrix $S(i, J, J)$ will appear in row L and column L of $[K]$.

Observe that owing to the symmetry of $[K]$ any 'symmetrical half' of it need only be assembled. Clearly, the explicit form of $[K]$ will be obtained by substituting into it the values of the $S(i, J, K)$'s. The above rules will now be illustrated by the following examples:

Example 1

Consider the beam of Fig. 2.2a; its discretized model has been redrawn in Fig. 3.1 in which for convenience we have identified the extremities by digits rather than by letters.

Fig. 3.1 A finite element model of the beam of Fig. 2.2a.

Focussing our attention to element 1, it follows from rule 1 that when extremity 1 visits extremity 2, $S(1, 1, 2)$ will appear (see Eq. 3.9a) in row 1 and column 2 of $[K]$, since in this case $i = 1$, $J = 1$, $K = 2$, $L = 1$ and $M = 2$. Also, according to rule 2, when extremity 1 visits itself, $S(1, 1, 1)$ will appear in row 1 and column 1 since now $i = 1$, $J = 1$ and $L = 1$.

Similarly, when extremity 2 visits extremity 1, we have $i = 1$, $J = 2$, $K = 1$, $L = 2$ and $M = 1$; as a result $S(1, 2, 1)$ appears in row 2 and column 1. Also, when extremity 2 visits itself, $i = 1$, $J = 2$ and $L = 2$ and consequently $S(1, 2, 2)$ appears in row 2 column 2.

According to our assumption the extremities of element 1 do not visit any other extremity. Element 1 does not therefore contribute anything else to $[K]$. This process, when repeated for elements 2 and 3, will result in the $[K]$ matrix of Eq. (3.9a).

Example 2

Next consider the combination of elements shown in Fig. 3.2. According to rule 2,

when extremity 1 of element 1 visits itself, $i = 1$, $J = 1$ and $L = 6$; consequently $S(1, 1, 1)$ will appear (see Eq. 3.12) in row 6, column 6;

when extremity 1 of element 2 visits itself, $i = 2$, $J = 1$ and $L = 6$; consequently $S(2, 1, 1)$ will appear in row 6, column 6;

when extremity 4 of element 3 visits itself, $i = 3$, $J = 4$ and $L = 6$; consequently $S(3, 4, 4)$ will appear in row 6, column 6.

It is left to the reader to verify by using these rules that in the case of Fig. 3.2

$$[K] = \begin{array}{c} \\ 6 \\ \\ 7 \\ \\ 8 \end{array} \left[\begin{array}{lll} \quad 6 & \quad 7 & \quad 8 \\ S(1, 1, 1) + S(2, 1, 1) + S(3, 4, 4) & S(2, 1, 2) + S(3, 4, 3) & S(1, 1, 2) + S(2, 1, 3) \\ S(2, 2, 1) + S(3, 3, 4) & S(2, 2, 2) + S(3, 3, 3) + S(4, 1, 1) & S(2, 2, 3) + S(4, 1, 3) \\ S(1, 2, 1) + S(2, 3, 1) & S(2, 3, 2) + S(4, 3, 1) & S(1, 2, 2) + S(2, 3, 3) + S(4, 3, 3) \end{array} \right] \tag{3.12}$$

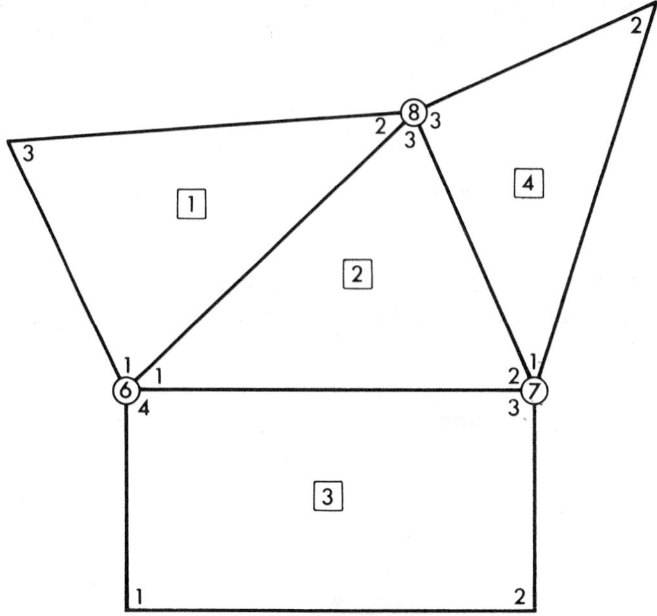

Fig. 3.2 A typical combination of triangular and rectangular elements.

The method given in this section together with the concept of the $S(i, J, K)$'s, constitutes a simple yet very satisfactory assembly procedure which is easily programmed for the computer. Indeed, we shall use it in Chapter 15.

The reader may find this rather mechanical method somewhat confusing at first. A little practice will, however, remedy this.

3.5 Further notes on the $S(i, J, K)$'s

In this section we shall discuss further the 'element submatrices' $S(i, J, K)$'s, which are perhaps the most important items in this chapter. In order to broaden the field, we shall discuss the $S(i, J, K)$'s of the triangular element of Fig. 11.1. The concept of these submatrices is the same in one, two and three dimensional finite elements.

It is clear here that the element's degree of freedom is 2; thus, each of its $S(i, J, K)$'s is a 2×2 matrix. Further, since the element has three extremities, its $[K_i]$ will contain a total of $3^2 = 9$ submatrices; also, J and K will vary as $J = 1, 2, 3$ and $K = 1, 2, 3$ (see section 3.2). Then, following a scheme similar

to that of Eq. (3.3), the $[K_i]$ of this element will be partitioned in a extremity-to-extremity basis and will be written in terms of its $S(i, J, K)$'s as

$$\begin{array}{l} K = 1, 2, 3 \longrightarrow \\ J = 1, 2, 3 \downarrow \end{array}$$

$$[K_i] = \begin{bmatrix} S(i, 1, 1) & S(i, 1, 2) & S(i, 1, 3) \\ S(i, 2, 1) & S(i, 2, 2) & S(i, 2, 3) \\ S(i, 3, 1) & S(i, 3, 2) & S(i, 3, 3) \end{bmatrix} \tag{3.13}$$

Clearly, the $S(i, J, K)$'s in Eq. (3.13) can be obtained explicitly in terms of the geometrical and material properties of element i, when its $[K_i]$ has also been found in those terms.

4 Boundary conditions

4.1 Rigid body displacement

The implications of 'boundary conditions' in the solution of a structural or continuum problem are intimately related to the concept of what is known as its 'rigid-body' displacements. We clarify this concept before undertaking a detailed discussion on boundary conditions.

Perhaps the simplest way to explain the somewhat difficult concept of 'rigid-body' displacement is to discuss it in connection with the single-element uniform bar 'structure', shown in Fig. 4.1. We assume that this 'structure', is completely free from all external forces except F_n and F_k, applied axially as shown. Since only axial forces are applied, the 'element stiffness matrix' of i can be obtained from Eq. (2.8) merely by deleting from it all rows and columns which refer to forces other than the axial forces. Thus, with $k_i = a_i E_i / h_i$,

$$[K_i] = k_i \begin{bmatrix} 1 & -1 \\ -1 & 1 \end{bmatrix}$$

As element i here is also the complete 'structure', $[K_i]$ is also its overall stiffness matrix. Furthermore, in this case

$$\{P\} = \begin{Bmatrix} -F_n \\ F_k \end{Bmatrix} \quad \text{and} \quad \{\delta\} = \begin{Bmatrix} u_n \\ u_k \end{Bmatrix}$$

Consequently, for this structure Eq. (3.9) becomes

$$k_i \begin{bmatrix} 1 & -1 \\ -1 & 1 \end{bmatrix} \begin{Bmatrix} u_n \\ u_k \end{Bmatrix} = \begin{Bmatrix} -F_n \\ F_k \end{Bmatrix} \tag{4.1}$$

Now consider the following cases:

(1) $F_k > F_n$

According to Newton's first law, the out-of-balance force $(F_k - F_n)$ will make the bar embark on a uniformly accelerated motion. Imagine that the bar's motion initiates at time $t = 0$. Then, in the absence of restraining forces, it is clear that during the time interval $(t_2 - t_1)$, where $t_2 > t_1 > 0$, all points on the bar would have undergone exactly the same displacement along x. Consequently, the bar does not suffer any elastic strain or stress. In structural and continuum mechanics a displacement of this type in which the body is not

Fig. 4.1 A single-element bar 'structure'. u_n and u_k are axial displacements at nodes n and k respectively.

strained or stressed is referred to as 'rigid body displacement'. The meaning of the term 'rigid' used here is opposite to that of 'elastic'. Thus, to generalize, a body is said to undergo a 'rigid body' displacement, which may be translational, rotational or mixed, when its displacement is not accompanied by elastic stresses or strains.

In the absence of restraints, u_n and u_k are the rigid body displacements of the nodes of the bar of Fig. 4.1. Since in this case all points on the bar have the same rigid body displacement along x, it follows that $u_n = u_k$. Let us now inquire if we could possibly determine u_n or u_k by solving Eq. (4.1). According to assumption 2 (section 2.1), Eq. (2.8) and hence Eq. (4.1) is valid for small displacements only. Let us therefore qualify u_n and u_k as being 'small'. It would soon become clear, however, that we are not able to solve Eq. (4.1) for u_n or u_k simply because this equation is not capable of a solution. Mathematically, this is due to the fact that the overall stiffness matrix

$$k_i \begin{bmatrix} 1 & -1 \\ -1 & 1 \end{bmatrix}$$

of the bar is 'singular', meaning that its inverse does not exist; as a result, Eq. (4.1) is indeterminate. We can therefore say that the rigid body displacement of this structure, and demonstrably of structures in general, is a consequence of the singularity of its overall stiffness matrix.

(2) $F_k = F_n$

The bar is now in equilibrium and extends by the amount $(u_k + u_n)$. Unlike in case (1) above, u_n and u_k are now elastic displacements resulting in elastic stresses and strains within the bar. This situation does not, however, remedy the singularity of the overall stiffness matrix and we are still unable to solve Eq. (4.1) for u_n and u_k. This then is an example in which the singularity of the overall stiffness matrix of the body does not lead to its rigid-body displacement.

Thus, we may draw this conclusion from the above cases: The rigid-body displacement of a body is due to the singularity of its overall stiffness matrix, although this singularity may not always lead to a rigid-body displacement.

4.2 Boundary conditions

The reason for the singularity of the overall stiffness matrix and the resulting insolubility of Eq. (4.1) can be attributed to the fact that we did not provide a

'datum', with respect to which u_n and u_k could be measured. Such a datum must be provided simply because displacements are relative quantities, and like all relative quantities, become meaningful only when they are measured with respect to a datum which has 'prescribed' or 'assigned' displacements.

Let us therefore assign a datum value, say zero, to either u_n or u_k; in particular, let $u_n = 0$. This condition is realized in practice when node n is restrained by fixing it to a rigid support as shown in Fig. 4.2. The magnitudes of F_n and F_k are now equal and we are able to solve Eq. (4.1) to obtain the elastic displacement u_k as

$$u_k = F_k/k_i$$

The condition, $u_n = 0$, that we have imposed here is a 'boundary condition' of the structure of Fig. 4.2. Thus, in elasto-mechanics, boundary conditions are constraints imposed on displacements* at a certain part or parts of the body. These constraints, in general, prevent the rigid-body displacements of the body and enable it to sustain externally applied forces.

Obviously, the boundary conditions provide the datum with respect to which the elastic displacements at various parts of the body will be measured. The overall stiffness matrix, $[K]$, of the body becomes 'non-singular' when the prescribed boundary conditions are incorporated into $[K]$. Then, when the external forces, $\{P\}$, are specified, Eq. (3.9) can be solved uniquely for the displacements $\{\delta\}$. Conversely, failure to satisfy the prescribed boundary conditions makes $[K]$ 'singular', as in cases 1 and 2 in section 4.1, and Eq. (3.9) indeterminate.

As a rule, a given structural or continuum problem has either 'natural' or 'artificial' boundary conditions. Consider, for instance, the problem of Fig. 5.1a in which the beam is fixed at each end; as a result, both vertical and rotational displacements at each end are zero and these in fact are the boundary conditions of this problem. To this structure these conditions are inherent as opposed to being contrived or assumed to make its $[K]$ matrix non-singular. We may therefore refer to them as 'natural' boundary conditions.

Next, consider the U-shaped prismatic beam structure of Fig. 4.3. Member i, which is initially too long by the amount e, is now forced into the open end of the

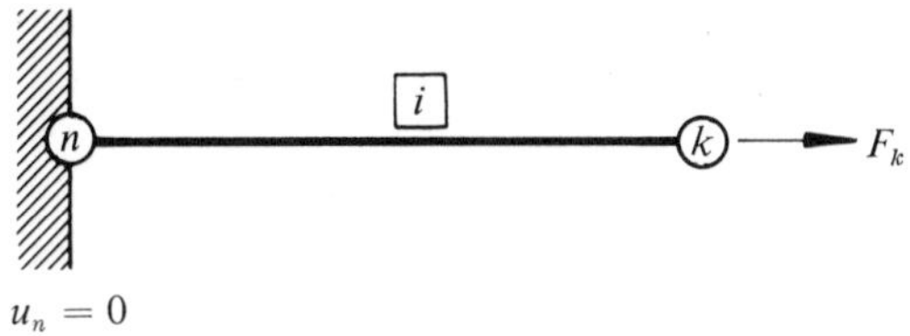

Fig. 4.2

* Sometimes, as in section 11.5, boundary conditions are also prescribed in terms of stress resultants. These, which have been excluded from present discussion, do not in general affect the properties of the overall stiffness matrix.

U as shown by the dotted lines. As a result, stresses are set up in the resulting O-shaped structure. In order to analyze this structure by finite elements, let us start by assembling its overall stiffness matrix, $[K]$. However, as the structure is not secured to any support, it does not have any 'natural' boundary condition (this is seen from the fact that when acted upon by an external force, this structure will displace as a rigid body). The $[K]$ matrix of this structure is therefore inherently singular. To make it non-singular, we must impose an 'artificial' datum displacement at any node. For instance, we may arbitrarily assume all displacements at the node at A to be zero. Then, an equation similar

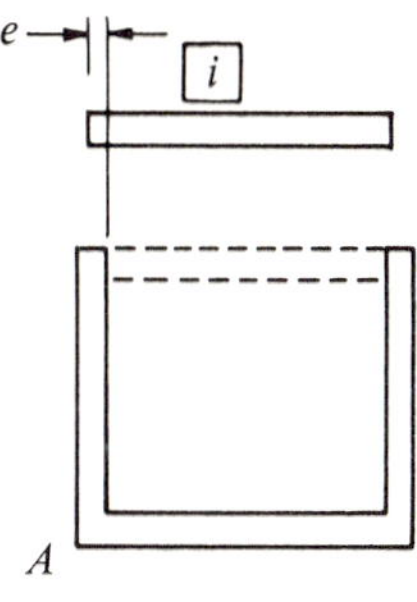

Fig. 4.3

to Eq. (3.9) can be solved to determine the displacements from which stresses can be calculated.* This then is an example of 'artificial' boundary conditions. A familiar example of this is provided by an aircraft in flight; if we regard the entire aircraft as an elastic structure, then, in the absence of natural boundary conditions, artificial boundary conditions would have to be imposed to make its overall stiffness matrix non-singular.

Most structural and continuum problems which occur in practice have natural boundary conditions. We shall therefore refer to them henceforth simply as 'boundary conditions'.

Since Eq. (3.9) relates the external nodal forces to the resulting nodal displacements, the boundary conditions need be satisfied only at the nodes representing the regions of the body at which these conditions are prescribed. The procedure for incorporating these conditions into Eq. (3.9) will be discussed in section 5.1.

Having accounted for all the prescribed boundary conditions, Eq. (3.9) can be solved uniquely, with a given $\{P\}$, for the unknown nodal displacements $\{\delta\}$. The stresses and strains within the body can then be determined from these displacements.

* A solution of this particular problem would require the introduction of 'transformation' matrices; see Chapter 6.

4.3 Hints for subdividing the body

In a given problem the size of the overall stiffness matrix and the degree of accuracy of solution are determined, to a large extent, by the manner in which the body is subdivided into a system of finite elements. A large measure of the analyst's skill and experience is called for in this matter. The general method of subdivision and its underlying logic will be discussed in sections 9.5 and 9.6. In the meantime, some general hints to this end are given below for the benefit of the beginner. These mainly apply to beam-type structures:

A node is usually provided at

(1) each support or point at which a displacement is prescribed, e.g., nodes 1 and 3 in Fig. 5.1b.

(2) Each point at which a concentrated external force is applied, e.g., node 2 in Fig. 5.1b (this is not necessary if 'equivalent force vectors' are used; see sections 9.3 and 9.7).

(3) Each junction where two or more members meet, e.g., node 2 in Fig. 6.5b, and

(4) each point at which stiffness changes suddenly, as in Fig. 4.4a. When stiffness changes gradually, as in Fig. 4.4b, a sufficient number of nodes must be provided, since results here converge to their exact values as the number of nodes is increased.* A similar situation obtains for curved structures, an example of which will be given in section 6.9.

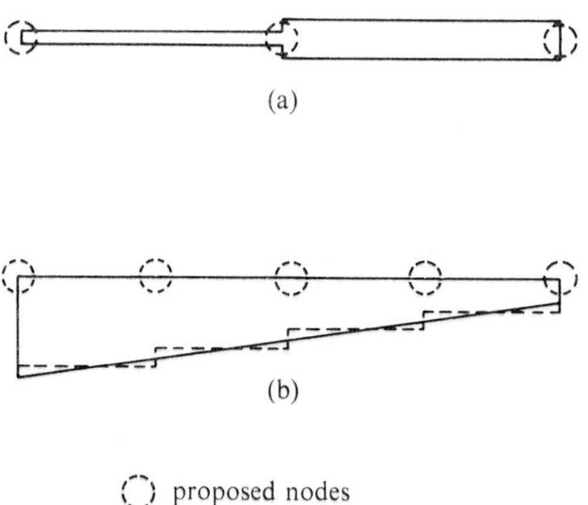

Fig. 4.4 (a) Sudden change of stiffness. (b) Gradual change of stiffness.

* If the variation of EI can be represented by a continuous function, then better results will be obtained by using the method outlined in section 9.8b.

5 Some worked out beam problems

Some beam structures are analyzed in this chapter to illustrate the salient points in the finite element method of solution developed in the previous chapters. The emphasis here is on the method of solution rather than on the size or complexity of the problem. The overall stiffness matrices of practical problems are usually large; consequently, their manual processing is wasteful and often impossible. However, as we shall see in Chapter 15, simple computer programs can be written to solve these large problems entirely automatically, provided sufficient computer store is available.

5.1 A uniform beam fixed at both ends

To begin with, consider the simple problem of Fig. 5.1a, a finite element subdivision of which is shown in Fig. 5.1b. Notice that as in Chapter 3, we have replaced in the finite element model the extremity notations a, b by the digits 1, 2 respectively, for the sake of convenience.

In the absence of axial and torsional forces, the degrees of freedom here is 2. The $S(i, J, K)$'s of the elements of this beam can therefore be obtained merely by excluding the axial and torsional modes from Eqs. (3.4); thus

$$S(i, 1, 1) = (E_i I_i / h_i^3) \begin{bmatrix} 12 & 6h_i \\ 6h_i & 4h_i^2 \end{bmatrix} \tag{5.1a}$$

$$S(i, 1, 2) = (E_i I_i / h_i^3) \begin{bmatrix} -12 & 6h_i \\ -6h_i & 2h_i^2 \end{bmatrix} \tag{5.1b}$$

$$S(i, 2, 2) = (E_i I_i / h_i^3) \begin{bmatrix} 12 & -6h_i \\ -6h_i & 4h_i^2 \end{bmatrix} \tag{5.1c}$$

and of course $S(i, 1, 2)$ and $S(i, 2, 1)$ are transposes of each other. The overall stiffness matrix for the model of Fig. 5.1b is now prepared by the method of section 3.4 to give

$$\begin{array}{cc} & \begin{array}{ccc} \quad 1 \quad & \quad\quad 2 \quad\quad & \quad 3 \quad \end{array} \\ [K] = \begin{array}{c} 1 \\ 2 \\ 3 \end{array} & \begin{bmatrix} S(1, 1, 1) & S(1, 1, 2) & 0 \\ S(1, 2, 1) & S(1, 2, 2) + S(2, 1, 1) & S(2, 1, 2) \\ 0 & S(2, 2, 1) & S(2, 2, 2) \end{bmatrix} \end{array} \tag{5.2}$$

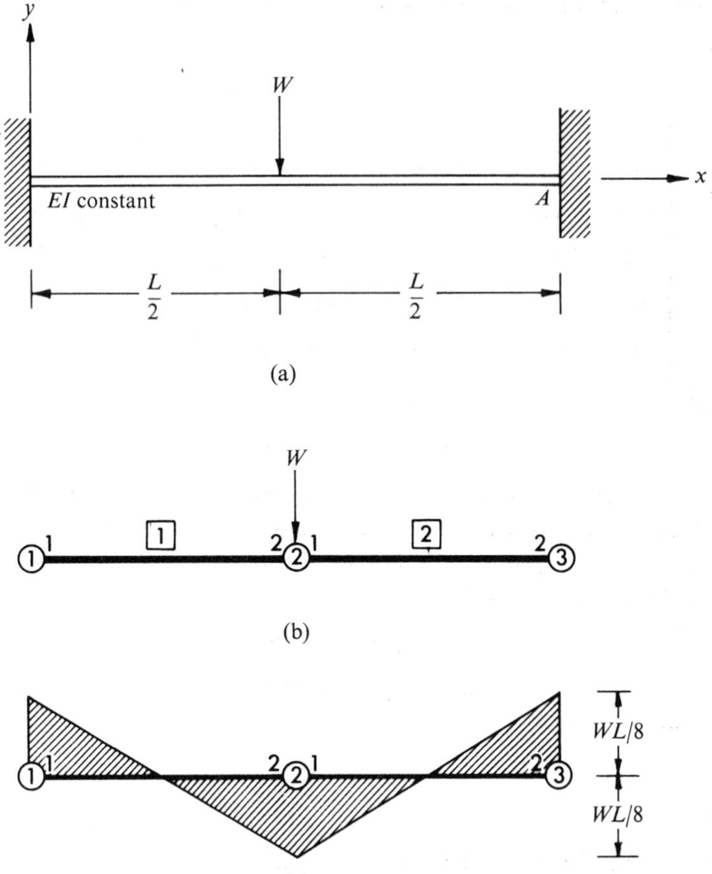

Fig. 5.1 (a) A uniform beam fixed at both ends. (b) A 2-element model of (a). (c) Computed moments in (a) obtained by using model (b).

As the total number of nodes here is 3, $N = 3$ and Eqs. (3.10) and (3.11) become

$$\{P\} = \{Q_1 M_1 \quad Q_2 M_2 \quad Q_3 M_3\}^{\mathrm{T}}$$

and

$$\{\delta\} = \{v_1 \theta_1 \quad v_2 \theta_2 \quad v_3 \theta_3\}^{\mathrm{T}}$$

respectively. Also, as the only non-zero external force here is Q_2, which is equal to $-W$, $\{P\}$ above becomes

$$\{P\} = \{0 \quad 0 \quad -W \quad 0 \quad 0 \quad 0\}^{\mathrm{T}}$$

For each element, its $S(i, J, K)$'s are now calculated from Eqs. (5.1) and substituted into Eq. (5.2). The resulting $[K]$ matrix, together with $\{P\}$ and $\{\delta\}$ is now substituted into Eq. (3.9), upon which Eq. (5.3) emerges. The rows

and columns of $[K]$ in Eq. (5.3) have been numbered for easy reference.

$$
\begin{array}{c}
\text{rows} \\ \downarrow \\ 1 \\ 2 \\ 3 \\ 4 \\ 5 \\ 6 \\ \text{columns} \rightarrow
\end{array}
\quad
q\begin{bmatrix}
12 & 6h & -12 & 6h & 0 & 0 \\
6h & 4h^2 & -6h & 2h^2 & 0 & 0 \\
-12 & -6h & 24 & 0 & -12 & 6h \\
6h & 2h^2 & 0 & 8h^2 & -6h & 2h^2 \\
0 & 0 & -12 & -6h & 12 & -6h \\
0 & 0 & 6h & 2h^2 & -6h & 4h^2
\end{bmatrix}
\begin{Bmatrix} v_1 \\ \theta_1 \\ v_2 \\ \theta_2 \\ v_3 \\ \theta_3 \end{Bmatrix}
=
\begin{Bmatrix} 0 \\ 0 \\ -W \\ 0 \\ 0 \\ 0 \end{Bmatrix}
\tag{5.3}
$$

$$\text{columns} \rightarrow \; 1 \quad 2 \quad 3 \quad 4 \quad 5 \quad 6$$

$$q = EI/h^3, \qquad h = L/2$$

Owing to its symmetry, in a given problem only any symmetrical half of $[K]$ need be assembled. $[K]$ is by nature 'singular' and will remain so until we introduce into it the prescribed boundary conditions (see section 4.2); in this problem these conditions are:

$$v_1 = \theta_1 = v_3 = \theta_3 = 0 \tag{5.4}$$

In finite element problems the $[K]$ matrix can be made non-singular by either of the two following methods:

Method 1

If displacement $\bar{U}$ is prescribed at a node, then replace by unity the diagonal term in the row referring to $\bar{U}$ and set every off-diagonal term of this row equal to zero. Also, replace the corresponding row of $\{P\}$ by $\bar{U}$.

If $[K]$ has a scalar multiplier, such as q in Eq. (5.3), then the diagonal term referring to $\bar{U}$ must be replaced by $1/q$ instead of unity.

Consider the condition $v_1 = 0$ of Eq. (5.4). Applying the above method, this condition will be accounted for when we (a) replace the diagonal term in row 1 of Eq. (5.3) by $1/q$ and every off-diagonal term in this row by zero, and (b) replace row 1 of $\{P\}$ by zero, since $\bar{U} = v_1 = 0$ (in this case row 1 of $\{P\}$ is already zero).

When Eq. (5.3) is modified by repeating this process for all the displacements of Eq. (5.4), it becomes non-singular; we can then solve it uniquely for v_2 and θ_2. This method, however, is wasteful since the storage of $[K]$ in the computer, in this problem for example, requires 36 'words' which compare with only 4 that would be needed in method 2 outlined below. Method 1 should therefore be used only for those rows which refer to non-zero prescribed displacements.

Method 2

Exclude from $[K]$ the rows and columns which refer to zero prescribed displacements and modify $\{P\}$ so that it is conformable with $[K]$. [This is justified as follows: let the displacement v at node k be prescribed as zero. Then, according to method 1, the equation represented by the row of v_k is

$$v_k = 0$$

As the right hand side is zero, this row does not affect any other displacement. It is pointless therefore to include this row and its associated column in Eq. (3.9).]

For the problem of Fig. 5.1a, the rows 1, 2, 5 and 6 and the corresponding columns are thus excluded from Eq. (5.3) by virtue of Eq. (5.4). Eq. (5.3) now becomes

$$\begin{bmatrix} 24 & 0 \\ 0 & 8h^2 \end{bmatrix} \begin{Bmatrix} v_2 \\ \theta_2 \end{Bmatrix} = \frac{h^3}{EI} \begin{Bmatrix} -W \\ 0 \end{Bmatrix} \tag{5.5}$$

Clearly, only 4 'words' are now required to store the reduced $[K]$ of Eq. (5.5). For an average machine the saving in storage space achieved for this particular problem is of no real consequence. However, in large problems, particularly continuum problems, the corresponding saving will be considerable. This method ought therefore to be used wherever possible. A manual solution of Eq. (5.5) gives the central displacements of the beam as

$$v_2 = -\frac{WL^3}{192EI} \quad \text{and} \quad \theta_2 = 0$$

which are in fact the exact results.

Next, let us solve the problem of Fig. 5.1a when the end A is given an initial rotation of $-\theta'$, so that the prescribed boundary conditions are now

$$v_1 = \theta_1 = v_3 = 0 \quad \text{and} \quad \theta_3 = -\theta'$$

According to method 1, the diagonal term in row 6 is replaced by (h^3/EI) and each off-diagonal term in this row by zero. Also, the 6-th. row of $\{P\}$ is replaced by $-\theta'$. Furthermore, following method 2, the rows of v_1, θ_1 and v_3 and the respective columns are deleted from $[K]$, thus reducing Eq. (5.3) to

$$\begin{bmatrix} 24 & 0 & 6h \\ 0 & 8h^2 & 2h^2 \\ 0 & 0 & h^3/EI \end{bmatrix} \begin{Bmatrix} v_2 \\ \theta_2 \\ \theta_3 \end{Bmatrix} = \frac{h^3}{EI} \begin{Bmatrix} -W \\ 0. \\ -\theta' \end{Bmatrix}$$

A manual solution of the above equation gives

$$v_2 = \left(\frac{L\theta'}{8} - \frac{WL^3}{192EI}\right) \quad \text{and} \quad \theta_2 = \frac{\theta'}{4}$$

This simple example illustrates the treatment of non-zero prescribed displacements by using method 1.

Calculation of internal forces

Once the nodal displacements have been computed, the internal forces in individual elements representing the structure can be found as follows:

As we saw in section 3.1, the displacements of an extremity must be equal to those of the node to which it is attached. Thus, in the case of Fig. 5.2,

$$\{\delta_{ai}\} = \{\delta_n\} \quad \text{and} \quad \{\delta_{bi}\} = \{\delta_k\}$$

Fig. 5.2 A uniform beam element.

Then, using the notation of section 2.7, we can write

$$\{\bar{\delta}\} = \begin{Bmatrix} \{\delta_{ai}\} \\ \{\delta_{bi}\} \end{Bmatrix} = \begin{Bmatrix} \{\delta_n\} \\ \{\delta_k\} \end{Bmatrix}$$

From Eq. (2.6b) and the above equation it follows therefore that the internal forces $\{\bar{P}\}$ within element i can be obtained from

$$\{\bar{P}\} = [K_i] \begin{Bmatrix} \{\delta_n\} \\ \{\delta_k\} \end{Bmatrix} \tag{5.6}$$

once we have calculated the nodal displacements $\{\delta_n\}$ and $\{\delta_k\}$.

As an example, let us now calculate the internal forces in element 1 ($i = 1$) of Fig. 5.1b with the boundary conditions of Eq. (5.4). Since the torsional and axial modes are absent, we have on setting $n = 1$ and $k = 2$

$$\begin{Bmatrix} \{\delta_1\} \\ \{\delta_2\} \end{Bmatrix} = \begin{Bmatrix} v_1 \\ \theta_1 \\ v_2 \\ \theta_2 \end{Bmatrix} = \begin{Bmatrix} 0 \\ 0 \\ -\dfrac{WL^3}{192EI} \\ 0 \end{Bmatrix}$$

The stiffness matrix for this element is

$$[K_1] = \begin{bmatrix} S(1,1,1) & S(1,1,2) \\ S(1,2,1) & S(1,2,2) \end{bmatrix}$$

in which the $S(1, J, K)$'s are obtained by setting $i = 1$ in Eqs. (5.1). It is now a simple matter to calculate the internal forces from Eq. (5.6). Fig. 5.1c shows the

internal moments of the beam found in this way. In practice, the calculation of internal forces, like all other computational aspects, will be left to the computer.

In the simple problem of Fig. 5.1 a prior knowledge of the variation of bending moment along the beam was used to draw the bending moment diagram, merely by calculating the internal moments at a single 'internal' node, namely node 2. Obviously, several such internal nodes would have to be provided when the variation of internal forces cannot be anticipated.

5.2 A uniform cantilever carrying a uniformly distributed load

A point load such as $-W$ in Fig. 5.1a can be directly substituted into $\{P\}$, since by definition all the components of $\{P\}$ are externally applied concentrated loads. The question now is: how could we introduce into $\{P\}$ a distributed load such as the one shown in Fig. 5.3a?

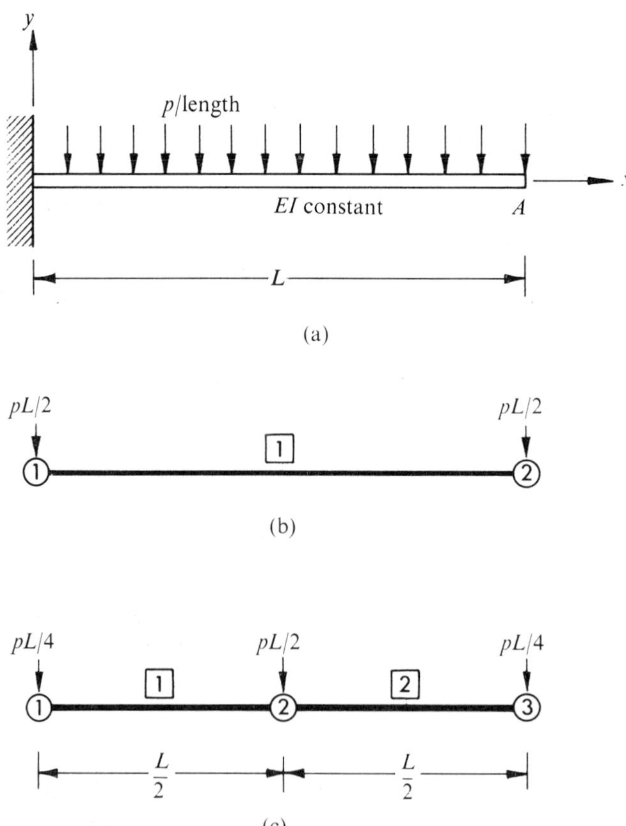

Fig. 5.3 (a) A uniform cantilever carrying a U.D.L. of intensity p. (b) A single element approximation of (a). (c) A two-element approximation of (a).

Let us first approximate this cantilever by a single element, and purely as a guess, assume that half the total load on it acts at each node as a concentrated vertical load as shown in Fig. 5.3b. Following method 2 of section 5.1, it is easy to show that in this case Eq. (3.9) takes the form

$$\begin{bmatrix} 12 & -6L \\ -6L & 4L^2 \end{bmatrix} \begin{Bmatrix} v_2 \\ \theta_2 \end{Bmatrix} = \frac{L^3}{EI} \begin{Bmatrix} -\dfrac{pL}{2} \\ 0 \end{Bmatrix} \tag{5.7}$$

Solving Eq. (5.7), the displacements of A are found to be

$$v_2 = -\frac{pL^4}{6EI} \quad \text{and} \quad \theta_2 = -\frac{pL^3}{4EI}$$

These results are, however, grossly in error, as the corresponding exact displacements are

$$-\frac{pL^4}{8EI} \quad \text{and} \quad -\frac{pL^3}{6EI}$$

respectively. When 2 elements are used, as in Fig. 5.3c, the accuracy improves considerably as Table 5.1 shows; it also shows that the results improve and converge to their exact values as we go on increasing the number of elements representing the beam.

This, however, is not satisfactory, as an accurate analysis would require structural members carrying distributed loads to be subdivided into a large number of elements, which would increase the size of $[K]$ considerably.

Table 5.1 Convergence of displacements at A (Fig. 5.3a)

Number of elements	v_A*	% error	θ_A†	% error
1	0·166 667	33·33	0·250 000	50·00
2	0·135 416	8·33	0·187 500	12·50
3	0·129 629	3·70	0·175 925	6·40
5	0·126 667	1·34	0·170 000	2·00
10	0·125 417	0·33	0·167 000	0·20

* Multiplier $= -\dfrac{pL^4}{EI}$; † Multiplier $= -\dfrac{pL^3}{EI}$

Let us therefore inquire as to how we could modify $\{P\}$, so that *exact* results are obtained when the beam of Fig. 5.3a is approximated by a *single* element, as in Fig. 5.3b. Treating the load vector as the unknown, and on substituting the exact displacements of A into Eq. (5.7), we have

$$\begin{bmatrix} 12 & -6L \\ -6L & 4L^2 \end{bmatrix} \begin{Bmatrix} -\dfrac{pL^4}{8EI} \\ -\dfrac{pL^3}{6EI} \end{Bmatrix} = \frac{L^3}{EI} \begin{Bmatrix} Q_2 \\ M_2 \end{Bmatrix}$$

the solution of which gives

$$\begin{Bmatrix} Q_2 \\ M_2 \end{Bmatrix} = \begin{Bmatrix} -\dfrac{pL}{2} \\ \dfrac{pL^2}{12} \end{Bmatrix}$$

This result shows that the cantilever can be analysed exactly with a single element approximation, if in addition to the point load $-pL/2$, we also apply the moment $pL^2/12$ at node 2. Releasing node 1 and fixing node 2, we can show in this way that

$$\begin{Bmatrix} Q_1 \\ M_1 \end{Bmatrix} = \begin{Bmatrix} -pL/2 \\ -pL^2/12 \end{Bmatrix}$$

Then, for the two degree-of-freedom element of Fig. 5.4a, we can define the so called 'equivalent force vector', $\{P_e\}$, as

$$\{P_e\} = \begin{Bmatrix} Q_n \\ M_n \\ Q_k \\ M_k \end{Bmatrix} = \begin{Bmatrix} ph/2 \\ ph^2/12 \\ ph/2 \\ -ph^2/12 \end{Bmatrix} \tag{5.8}$$

Fig. 5.4b displays the definition of $\{P_e\}$. Clearly, $\{P_e\}$ can be used to represent members carrying uniformly distributed forces by a single element. A typical application of $\{P_e\}$ will be given in the next section. In Chapter 9 we shall derive a general procedure for calculating $\{P_e\}$ of any finite element.

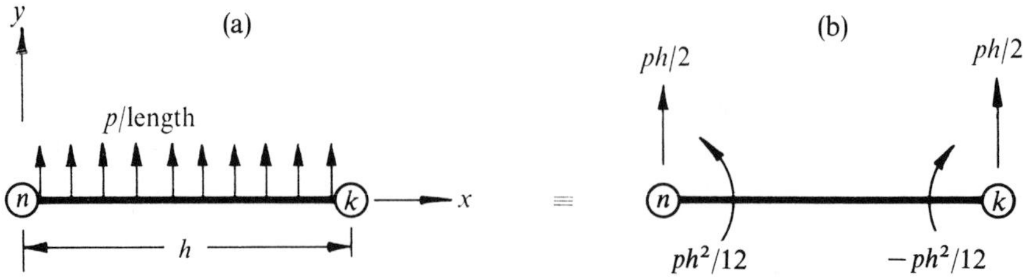

Fig. 5.4 (a) A beam element carrying a U.D.L. (b) Pictorial definition of $\{P_e\}$ for (a).

5.3 Some examples

Example 1

Problems in which the geometry, the loads and the boundary conditions are symmetrical about one or more axes, the overall stiffness matrix need only be assembled for any symmetrical region of the body. Clearly, this will drastically reduce the size of the $[K]$ matrix.

Thus, in the problem of Fig. 5.5a we need assemble the $[K]$ matrix for either of the symmetrical halves of the beam, as shown in Fig. 5.5b. However, we must introduce the additional boundary condition $\theta_3 = 0$, which is consistent with the deformed shape of the loaded beam. The prescribed boundary conditions of this problem are therefore

$$v_1 = \theta_1 = \theta_3 = 0.$$

Now, deleting from Eq. (2.8) the rows and columns corresponding to the axial and torsional modes, we can show that for Fig. 5.5b

$$[K_1] = [K_2] = \frac{EI}{27}\begin{bmatrix} 12 & 18 & -12 & 18 \\ 18 & 36 & -18 & 18 \\ -12 & -18 & 12 & -18 \\ 18 & 18 & -18 & 36 \end{bmatrix} \tag{5.9}$$

Since all displacements vanish at node 1, this node will not be included in $[K]$ (method 2, section 5.1). Then, using the methods of Chapter 3 we can show that for Fig. 5.5b

$$[K] = \begin{matrix} & 2 & 3 \\ 2 & \left[S(1, 2, 2) + S(2, 1, 1) \right. & \left. S(2, 1, 2) \right] \\ 3 & \left[S(2, 2, 1) \right. & \left. S(2, 2, 2) \right] \end{matrix} \tag{5.10}$$

or, partitioning the square matrix of Eq. (5.9) into $S(i, J, K)$'s and making appropriate substitutions, we have

$$[K] = \frac{EI}{27}\begin{bmatrix} 24 & 0 & -12 & 18 \\ 0 & 72 & -18 & 18 \\ -12 & -18 & 12 & -18 \\ 18 & 18 & -18 & 36 \end{bmatrix} \tag{5.11}$$

The uniformly distributed load will now be replaced by $\{P_e\}$; setting $n = 1$

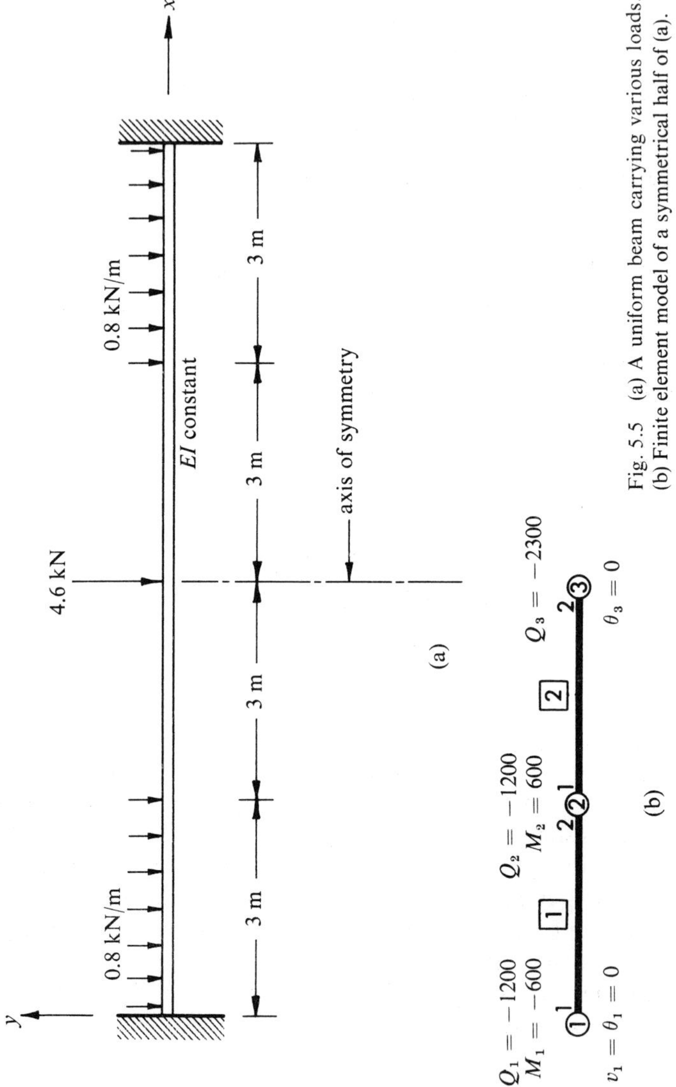

Fig. 5.5 (a) A uniform beam carrying various loads. (b) Finite element model of a symmetrical half of (a).

and $k = 2, p = -800$ N/m and $h = 3$ m into Eq. (5.8) we have

$$\{P_e\} = \begin{Bmatrix} Q_1 \\ M_1 \\ Q_2 \\ M_2 \end{Bmatrix} = \begin{Bmatrix} -1200 \\ -600 \\ -1200 \\ 600 \end{Bmatrix}$$

in which the quantities are in SI units (see Appendix 1 for conversion to the F.P.S units). Consequently, the nodal forces in the case of Fig. 5.5b are

$$Q_1 = Q_2 = -1200 \text{ N}, \qquad M_1 = -600 \text{ Nm}, \qquad M_2 = 600 \text{ Nm}$$

and

$$Q_3 = -2300 \text{ N}$$

Therefore, since we have excluded node 1 the required external force vector is

$$\{P\} = \{Q_2 \quad M_2 \quad Q_3 \quad M_3\}^{\mathrm{T}} = \{-1200 \quad 600 \quad -2300 \quad 0\}^{\mathrm{T}} \tag{5.12}$$

Furthermore, since $\theta_3 = 0$, the row and column referring to θ_3 will also be excluded (method 2, section 5.1). Consequently, from Eqs. (3.9), (5.11) and (5.12) we have

$$\frac{EI}{27}\begin{bmatrix} 24 & 0 & -12 \\ 0 & 72 & -18 \\ -12 & -18 & 12 \end{bmatrix}\begin{Bmatrix} v_2 \\ \theta_2 \\ v_3 \end{Bmatrix} = \begin{Bmatrix} -1200 \\ 600 \\ -2300 \end{Bmatrix} \tag{5.13}$$

in which EI is in SI units. A solution of Eq. (5.13) gives

$$v_2 = -26100{\cdot}0/EI \text{ (m)}$$

$$\theta_2 = -12150{\cdot}0/EI \text{ (rad)}$$

$$v_3 = -49500{\cdot}0/EI \text{ (m)}$$

The internal forces within individual elements can now be found from Eq. (5.6). However, in the case of elements carrying uniformly distributed loads the approach is slightly different as explained below:

Although $\{P_e\}$ is treated as one, it is not strictly speaking an external force vector but merely an expedient for calculating accurately the displacements of structures carrying uniformly distributed loads. Consequently, in the case of elements subjected to uniformly distributed loads, $\{P_e\}$ must be subtracted from the internal forces given by Eq. (5.6) in order to arrive at the true internal forces; thus, in the case of such elements Eq. (5.6) becomes

$$\{\bar{P}\} = [K_i]\begin{Bmatrix} \{\delta_n\} \\ \{\delta_k\} \end{Bmatrix} - \{P_e\} \tag{5.14}$$

As an example of this consider element 1 of Fig. 5.5b; setting $n = 1$, $k = 2$ and $i = 1$ into Eq. (5.14) and using $[K_1]$ of Eq. (5.9), we can show that the true

internal forces within this element are

$$\begin{Bmatrix} Q_{a1} \\ M_{a1} \\ Q_{b1} \\ M_{b1} \end{Bmatrix} = [K_1] \begin{Bmatrix} v_1 \\ \theta_1 \\ v_2 \\ \theta_2 \end{Bmatrix} - \{P_e\}$$

$$= \frac{1}{27} \begin{bmatrix} 12 & 18 & -12 & 18 \\ 18 & 36 & -18 & 18 \\ -12 & -18 & 12 & -18 \\ 18 & 18 & -18 & 36 \end{bmatrix} \begin{Bmatrix} 0 \\ 0 \\ -26100 \\ -12150 \end{Bmatrix} - \begin{Bmatrix} -1200 \\ -600 \\ -1200 \\ 600 \end{Bmatrix}$$

$$= \begin{Bmatrix} 4700 \\ 9900 \\ -2300 \\ 600 \end{Bmatrix}$$

where all quantities are in SI units. Clearly, the bending moment at extremity 1 of this element is 9·9 kNm, etc.

Example 2

Consider the problem of Fig. 5.6a, whose finite element model is shown in Fig. 5.6b. Since all displacements vanish at the nodes 1 and 3, these nodes need not be included in $[K]$. Consequently, following the methods of Chapter 3 we can show that in the case of Fig. 5.6b

$$[K] = 2 \quad S(1, 2, 2) \overset{2}{+} S(2, 1, 1)$$

Setting $E_i I_i = 5EI$ and $h_i = 5$ into Eq. (5.1c) we have for element 1

$$S(1, 2, 2) = \frac{EI}{25} \begin{bmatrix} 12 & -30 \\ -30 & 100 \end{bmatrix}$$

Similarly, with $i = 2$, $h_i = 10$ and $E_i I_i = EI$ Eq. (5.1a) gives

$$S(2, 1, 1) = \frac{EI}{1000} \begin{bmatrix} 12 & 60 \\ 60 & 400 \end{bmatrix}$$

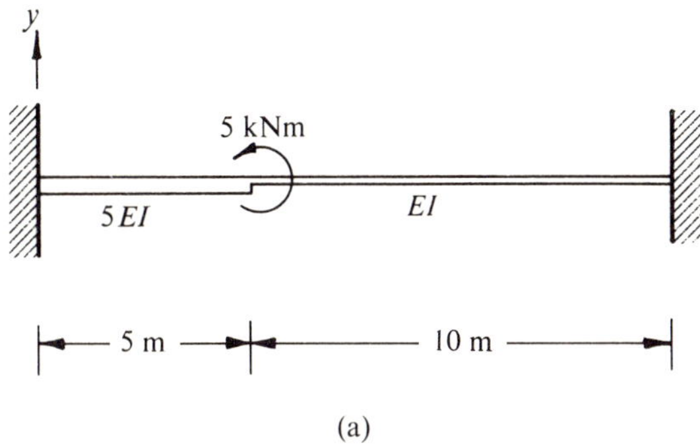

(a)

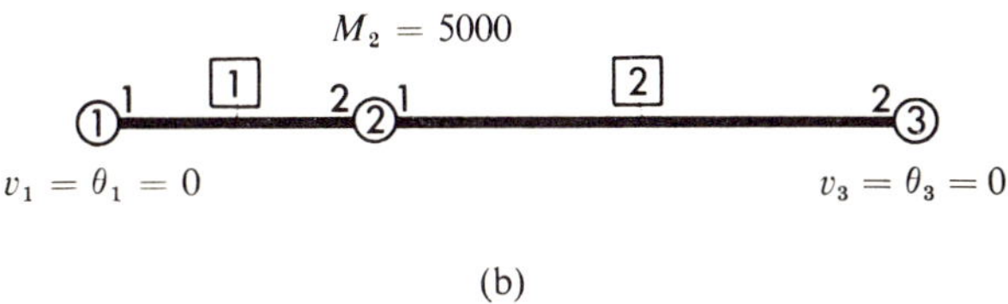

(b)

Fig. 5.6 (a) A nonuniform beam subjected to a moment. (b) Finite element model of (a).

Consequently,
$$[K] = \frac{EI}{1000}\begin{bmatrix} 492 & -1140 \\ -1140 & 4400 \end{bmatrix}$$

In this problem the only external force is $M_2 = 5$ kNm, and therefore
$$\{P\} = \{0 \quad 5000\}^T$$

As a result Eq. (3.9) becomes
$$EI\begin{bmatrix} 0{\cdot}492 & -1{\cdot}140 \\ -1{\cdot}140 & 4{\cdot}400 \end{bmatrix}\begin{Bmatrix} v_2 \\ \theta_2 \end{Bmatrix} = \begin{Bmatrix} 0 \\ 5000 \end{Bmatrix} \tag{5.15}$$

Solving Eq. (5.15) we get
$$v_2 = 6{\cdot}587/EI \text{ (m)}$$
$$\theta_2 = 2{\cdot}843/EI \text{ (rad)}$$

The internal forces can now be found from Eq. (5.6).

Finally, it may be remarked that in both the above examples the finite element results coincide with the respective exact results. In the case of beam structures this is to be generally expected.

6 The rigid-jointed frame

6.1 General

A rigid-jointed frame structure is made up of a number of beam members which are variously aligned either on a plane or in space, and are interconnected by rigid joints. When the members of such a frame are co-planar, the structure is called a plane frame, and a space frame when they are variously oriented in space. A 'rigid' joint is defined as one that is able to sustain both moments and forces. Consequently, frames such as these are designed to sustain all kinds of loads which may be applied at the joints and/or members of the frames themselves. The beam element of Chapter 2 can also be used to analyze these structures provided that we modify its $S(i, J, K)$'s to account for the different alignments of the members. This can be done, without difficulty, by using the so called 'transformation' matrix.

6.2 Concept of the transformation matrix

A consequence of the different orientations of the frame members is that we can no longer express the forces and displacements in all the elements representing the frame with respect to a single coordinate system, in the way we could in the previous chapters. Consider, for instance, the elements of Fig. 6.1. Treating each element individually, the forces and displacements of element 1 will be measured with respect to its 'local' coordinate axes, which is yax. Similarly, those of element 2 will be referred to its local axes, $y'ax'$. It is obvious, however, that any linear dimension will usually have unequal projections in the two local coordinate systems, as long as $\alpha \neq 0$. Therefore, in order to have a coherent formulation of problems of this type with inclined members, we must devise a so called 'global' coordinate system which is common to all finite elements into which the frame is subdivided, and with respect to which all forces and displacements would be measured. Suppose that a certain vector quantity, when measured in the local axes of element i, is found to be $\{B_i\}$. Also, let the measured value of the same quantity in the global axes be $\{B^g\}$. Then, these two values of the quantity would be related to each other as

$$\{B^g\} = [T_i]\{B_i\}$$

The matrix $[T_i]$ is called the 'transformation' matrix of element i, since it transforms the value of the quantity from the local to the global axes. $[T_i]$ depends, as we shall see, on the orientation of the local axes relative to the global axes. Thus, in Fig. 6.1, $[T_2]$ will transform any vector quantity (such

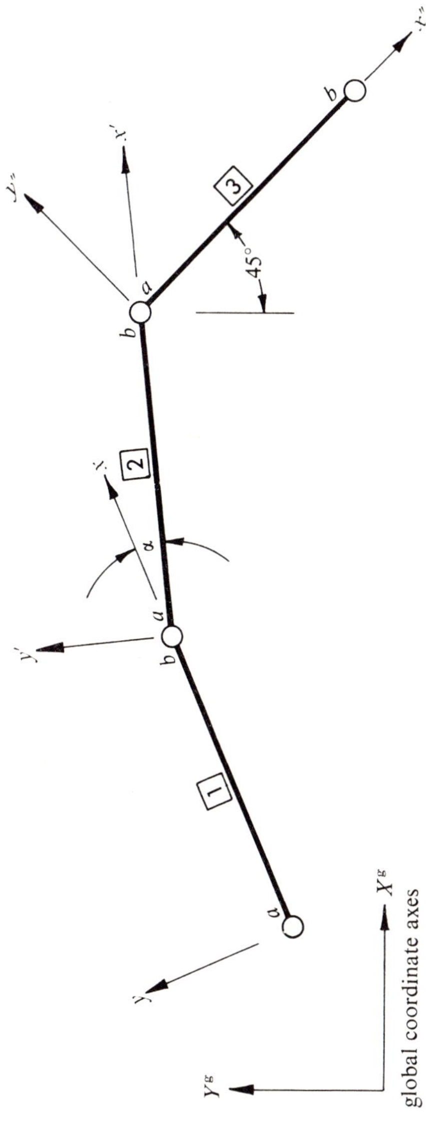

Fig. 6.1 Some inclined co-planar beam elements; yax, $y'ax'$ and $y''ax''$ are the 'local' axes of elements 1, 2 and 3 respectively.

as forces, displacements etc.) as measured in the local axes $y'ax'$, to its corresponding value in the global axes Y^g0X^g.

Here, as in the rest of this book, the superscript g stands for items which refer to the global axes. Thus, u_k^g and v_k^g, for example, denote the displacements of node k along X^g and Y^g respectively.

6.3 Derivation of the plane transformation matrix

Consider element i in Fig. 6.2 which belongs to a plane frame and has 3 degrees of freedom as in Fig. 2.2b (torsional forces will be considered in connection with space frames in which they are more relevant). Recalling that the superscript g denotes items in the global axes, it is clear that at extremity a, the nett global components of horizontal and vertical forces are

$$F_{ai}^g = F_{ai}\cos\alpha_i - Q_{ai}\sin\alpha_i$$

and
$$Q_{ai}^g = F_{ai}\sin\alpha_i + Q_{ai}\cos\alpha_i$$

respectively.

As usual, the symbols with only inferior suffixes denote quantities measured in the local axes yax. Also, notice that α_i is always measured anti-clockwise from the $+X^g$ axis, by imagining extremity a of the element to coincide with the origin of the global axes. Thus, in the case of element 3 in Fig. 6.1, for example, $\alpha_3 = 315°$.

Similarly, for extremity b the reader may verify that

$$F_{bi}^g = F_{bi}\cos\alpha_i - Q_{bi}\sin\alpha_i$$

and,
$$Q_{bi}^g = F_{bi}\sin\alpha_i + Q_{bi}\cos\alpha_i$$

Further, as moments and rotations in this case are the same in both local and global systems, we may write,

$$M_{ai}^g = M_{ai}$$

and,
$$M_{bi}^g = M_{bi}$$

Let us now put all the above equations in the matrix form as follows:

$$\begin{Bmatrix} F_{ai}^g \\ Q_{ai}^g \\ M_{ai}^g \\ \hline F_{bi}^g \\ Q_{bi}^g \\ M_{bi}^g \end{Bmatrix} = \left[\begin{array}{ccc|ccc} l_i & -m_i & 0 & 0 & 0 & 0 \\ m_i & l_i & 0 & 0 & 0 & 0 \\ 0 & 0 & 1 & 0 & 0 & 0 \\ \hline 0 & 0 & 0 & l_i & -m_i & 0 \\ 0 & 0 & 0 & m_i & l_i & 0 \\ 0 & 0 & 0 & 0 & 0 & 1 \end{array}\right] \begin{Bmatrix} F_{ai} \\ Q_{ai} \\ M_{ai} \\ \hline F_{bi} \\ Q_{bi} \\ M_{bi} \end{Bmatrix} = [T_i] \begin{Bmatrix} F_{ai} \\ Q_{ai} \\ M_{ai} \\ F_{bi} \\ Q_{bi} \\ M_{bi} \end{Bmatrix} \tag{6.1}$$

Clearly, the square matrix in Eq. (6.1) relates the extremity forces in the

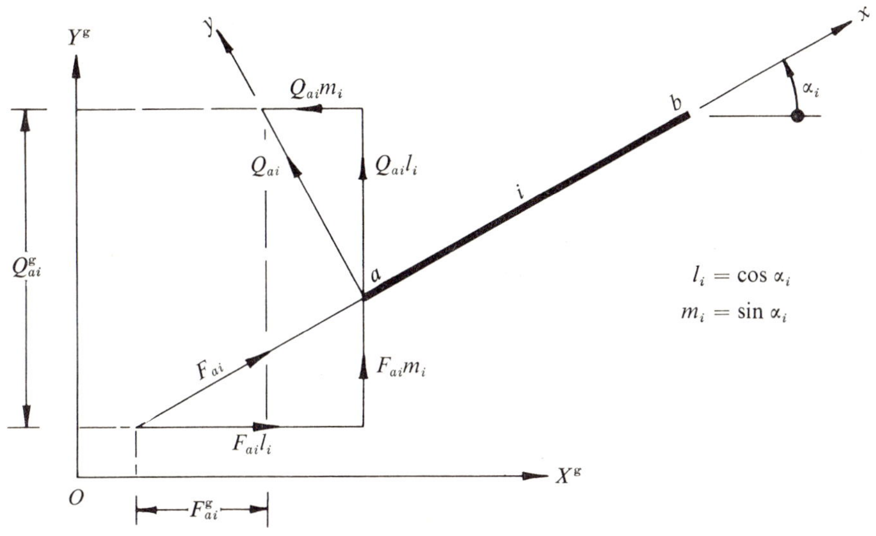

Fig. 6.2 Linear extremity loads in local and global systems.

local coordinates to those in the global system; by the definition of the last section, this matrix is therefore the required transformation matrix $[T_i]$ of element i (ignore the dotted lines for the time being). Obviously, for a given element the order of $[T_i]$ is equal to that of its $[K_i]$. An important property of this transformation matrix is that its transpose is also equal to its inverse; that is,

$$[T_i]^T = [T_i^{-1}] \tag{6.2}$$

Consequently, $[T_i]$ is said in matrix algebra to represent an 'orthogonal' transformation. Following an identical procedure, we can also show that

$$\begin{Bmatrix} u^g_{ai} \\ v^g_{ai} \\ \theta^g_{ai} \\ u^g_{bi} \\ v^g_{bi} \\ \theta^g_{bi} \end{Bmatrix} = [T_i] \begin{Bmatrix} u_{ai} \\ v_{ai} \\ \theta_{ai} \\ u_{bi} \\ v_{bi} \\ \theta_{bi} \end{Bmatrix} \tag{6.3}$$

The left hand vector of Eq. (6.3) obviously denotes the element's extremity displacements referred to the global axes, while the right hand vector lists the corresponding displacements in its local axes. Eqs. (6.1) and (6.3) show that

both forces and displacements of the element are transformed by the same matrix, as we might have anticipated.

6.4 The 'global' stiffness matrix of a plane element

By derivation, the $[K_i]$ matrix of element i refers to its local axes. Let us now determine how the $[K_i]$ of the element of section 6.3 transforms into the global axes.

Premultiplying both sides of Eq. (6.3) by $[T_i^{-1}]$ and using Eq. (6.2), we can write Eq. (6.3) also as

$$\begin{Bmatrix} u_{ai} \\ v_{ai} \\ \theta_{ai} \\ u_{bi} \\ v_{bi} \\ \theta_{bi} \end{Bmatrix} = [T_i]^{\mathrm{T}} \begin{Bmatrix} u^{\mathrm{g}}_{ai} \\ v^{\mathrm{g}}_{ai} \\ \theta^{\mathrm{g}}_{ai} \\ u^{\mathrm{g}}_{bi} \\ v^{\mathrm{g}}_{bi} \\ \theta^{\mathrm{g}}_{bi} \end{Bmatrix} \tag{6.4}$$

As we have excluded the torsional mode, the element stiffness matrix $[K_i]$ can be obtained from Eq. (2.8) merely by deleting from it the rows and columns which refer to torsional forces and displacements. Eq. (2.8) will then read

$$\begin{Bmatrix} F_{ai} \\ Q_{ai} \\ M_{ai} \\ F_{bi} \\ Q_{bi} \\ M_{bi} \end{Bmatrix} = [K_i] \begin{Bmatrix} u_{ai} \\ v_{ai} \\ \theta_{ai} \\ u_{bi} \\ v_{bi} \\ \theta_{bi} \end{Bmatrix} \tag{6.5}$$

in which $[K_i]$ is a 6×6 matrix. Premultiplication of both sides of Eq. (6.5) by $[T_i]$ leads to

$$[T_i] \begin{Bmatrix} F_{ai} \\ Q_{ai} \\ M_{ai} \\ F_{bi} \\ Q_{bi} \\ M_{bi} \end{Bmatrix} = [T_i K_i] \begin{Bmatrix} u_{ai} \\ v_{ai} \\ \theta_{ai} \\ u_{bi} \\ v_{bi} \\ \theta_{bi} \end{Bmatrix} \tag{6.6}$$

Making appropriate substitutions from Eqs. (6.1) and (6.4) into Eq. (6.6), we can then show that Eq. (6.6) can also be written as

$$\begin{Bmatrix} F^g_{ai} \\ Q^g_{ai} \\ M^g_{ai} \\ F^g_{bi} \\ Q^g_{bi} \\ M^g_{bi} \end{Bmatrix} = [T_i K_i T_i^T] \begin{Bmatrix} u^g_{ai} \\ v^g_{ai} \\ \theta^g_{ai} \\ u^g_{bi} \\ v^g_{bi} \\ \theta^g_{bi} \end{Bmatrix} \tag{6.7}$$

Clearly, the matrix product $[T_i K_i T_i^T]$ in Eq. (6.7) relates the global extremity forces of element i to its global extremity displacements. The matrix

$$[K^g_i] = [T_i K_i T_i^T]$$

will therefore be defined as the 'transformed' or 'global' stiffness matrix of element i. By contrast, $[K_i]$ relates the extremity forces and displacements of i when they are measured in the local axes of i. A multiplication will show that $[K^g_i]$ is always a square symmetric matrix, as expected.

6.5 The global submatrices of a plane frame element

In plane frames, as in the straight beams of the previous chapters, the steps leading up to the final solution of the problem are considerably simplified by splitting the element stiffness matrix into its submatrices. Let $[t_i]$ be a submatrix of $[T_i]$ such that*

$$[T_i] = \begin{bmatrix} t_i & 0 \\ 0 & t_i \end{bmatrix} \tag{6.8}$$

where the 0's denote null (zero) matrices. Then, comparing Eqs. (6.1) and (6.8), we have

$$[t_i] = \begin{bmatrix} l_i & -m_i & 0 \\ m_i & l_i & 0 \\ 0 & 0 & 1 \end{bmatrix} \tag{6.9}$$

Substituting for $[T_i]$ and $[K_i]$ from Eqs. (6.8) and (3.3) respectively, we have by definition

$$[K^g_i] = [T_i K_i T_i^T] = \begin{bmatrix} t_i & 0 \\ 0 & t_i \end{bmatrix} \begin{bmatrix} S(i,1,1) & S(i,1,2) \\ S(i,2,1) & S(i,2,2) \end{bmatrix} \begin{bmatrix} t_i^T & 0 \\ 0 & t_i^T \end{bmatrix}$$

$$= \begin{bmatrix} S^g(i,1,1) & S^g(i,1,2) \\ S^g(i,2,1) & S^g(i,2,2) \end{bmatrix} \tag{6.10}$$

* The t_i's and 0's are formed by partitioning $[T_i]$ in Eq. (6.1) along the dotted lines.

It may be verified by multiplication that in Eq. (6.10)

$$S^g(i, J, K) = (t_i \,.\, S(i, J, K) \,.\, t_i^T), \qquad J = 1, 2 \quad \text{and} \quad K = 1, 2 \quad (6.11)$$

A comparison between Eqs. (6.10) and (3.3) shows that the global overall stiffness matrix $[K^g]$ of the whole structure will also be assembled by the methods of Chapter 3. However, instead of assembling the $S(i, J, K)$'s as we did in Chapter 3 we now assemble the $S^g(i, J, K)$'s of elements. Obviously, the externally applied nodal forces and the resulting nodal displacements must now also refer to the global axes. Accordingly, if the frame is represented by a total of N nodes, then the listings of $\{P^g\}$ and $\{\delta^g\}$ will be as follows:

$$\{P^g\} = \{F_1^g\ Q_1^g\ M_1^g \quad F_2^g\ Q_2^g\ M_2^g \cdots F_N^g\ Q_N^g\ M_N^g\}^T \qquad (6.12)$$

and

$$\{\delta^g\} = \{u_1^g\ v_1^g\ \theta_1^g \quad u_2^g\ v_2^g\ \theta_2^g \cdots u_N^g\ v_N^g\ \theta_N^g\}^T \qquad (6.13)$$

Here F_k^g and Q_k^g are components of all linear external forces applied at node k, resolved along the X^g and Y^g axes respectively; M_k^g is the externally applied moment at node k (notice that in plane frames $M_k^g = M_k$). Then, for a rigid jointed plane frame and indeed for frame structures in general Eq. (3.9) will take the following form:

$$\{P^g\} = [K^g]\{\delta^g\} \qquad (6.14)$$

After inserting the prescribed boundary conditions into Eq. (6.14), it can be solved to give the global nodal displacements, $\{\delta^g\}$. A typical application of the above formulae will be given in section 6.8.

6.6 Rigid-jointed space frames

The procedure developed in the previous sections for plane frames can be extended to space frames without difficulty. The difference here, however, is that the element has 6 degrees-of-freedom, 3 of translation and 3 of rotation as shown in Fig. 6.3. Also, it is no longer necessary to distinguish between torsional and bending modes of loading as it was for the plane elements of Chapter 2, since for instance, M_{xk} in Fig. 6.3 which is a torsional load on i, is in fact a bending moment to an element with its axis along Oz.

[*Notation* Notice that different symbols are used for moments and rotations. M_{xai} and θ_{xai}, for instance, denote the internal moment and rotation respectively at extremity a of element i, both being on a plane normal to the x-axis. Similarly M_{zk} and θ_{zk} denote the (externally applied) moment and rotation respectively at node k, both being on a plane normal to the z-axis, etc. Also, I_{xi}, I_{yi} and I_{zi} denote the second moment of area of element i about the x, y and z axes respectively.]

It is more convenient in space frames to arrange the force and displacement vectors such that all the translational and rotational quantities are grouped together separately. The counterpart of Eq. (2.8) for the element of Fig. 6.3

thus becomes

$$\begin{Bmatrix} F_{ai} \\ Q_{ai} \\ H_{ai} \\ M_{xai} \\ M_{yai} \\ M_{zai} \\ F_{bi} \\ Q_{bi} \\ H_{bi} \\ M_{xbi} \\ M_{ybi} \\ M_{zbi} \end{Bmatrix} = [K_i] \begin{Bmatrix} u_{ai} \\ v_{ai} \\ w_{ai} \\ \theta_{xai} \\ \theta_{yai} \\ \theta_{zai} \\ u_{bi} \\ v_{bi} \\ w_{bi} \\ \theta_{xbi} \\ \theta_{ybi} \\ \theta_{zbi} \end{Bmatrix} \tag{6.15}$$

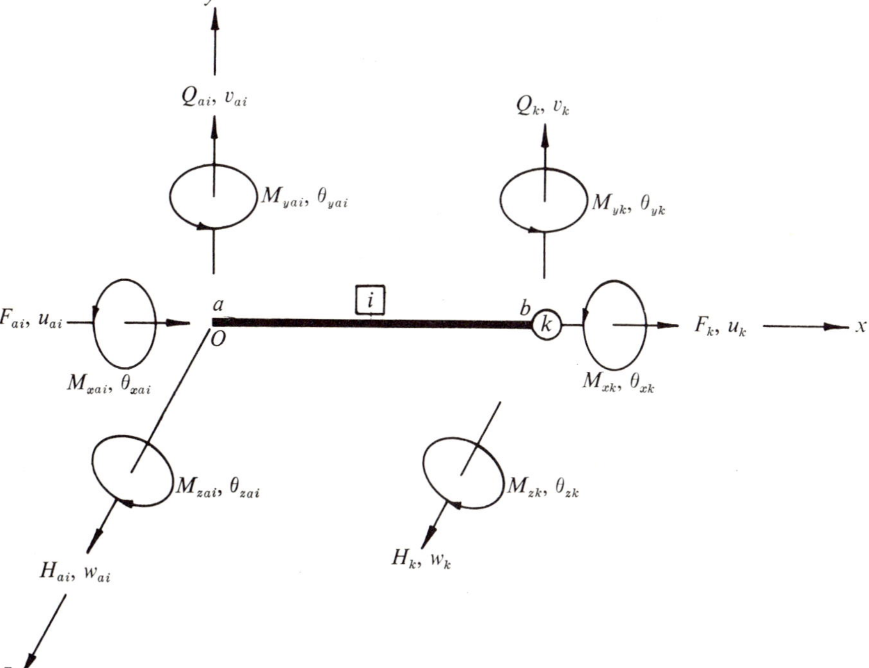

Fig. 6.3 A space frame element; forces and displacements refer to the local coordinates *oxyz*.

The element stiffness matrix $[K_i]$ above, which is with respect to the local coordinates of i as shown in Fig. 6.3, is obviously a 12×12 matrix. According to section 3.3, each of its $S(i, J, K)$'s is of size 6×6, and $J = 1, 2$ and $K = 1, 2$. The coefficients of $[K_i]$ and hence its $S(i, J, K)$'s can now be calculated by the method of Chapter 2, by taking into consideration the various orientations of different forces and displacements. The reader can verify that,

$$S(i,1,1) = E_i \begin{bmatrix} a_i/h_i & 0 & 0 & 0 & 0 & 0 \\ 0 & 12I_{zi}/h_i^3 & 0 & 0 & 0 & 6I_{zi}/h_i^2 \\ 0 & 0 & 12I_{yi}/h_i^3 & 0 & -6I_{yi}/h_i^2 & 0 \\ 0 & 0 & 0 & \beta_i/h_i & 0 & 0 \\ 0 & 0 & -6I_{yi}/h_i^2 & 0 & 4I_{yi}/h_i & 0 \\ 0 & 6I_{zi}/h_i^2 & 0 & 0 & 0 & 4I_{zi}/h_i \end{bmatrix} \tag{6.16a}$$

and,

$$S(i,1,2) = E_i \begin{bmatrix} -a_i/h_i & 0 & 0 & 0 & 0 & 0 \\ 0 & -12I_{zi}/h_i^3 & 0 & 0 & 0 & 6I_{zi}/h_i^2 \\ 0 & 0 & -12I_{yi}/h_i^3 & 0 & -6I_{yi}/h_i^2 & 0 \\ 0 & 0 & 0 & -\beta_i/h_i & 0 & 0 \\ 0 & 0 & 6I_{yi}/h_i^2 & 0 & 2I_{yi}/h_i & 0 \\ 0 & -6I_{zi}/h_i^2 & 0 & 0 & 0 & 2I_{zi}/h_i \end{bmatrix} \tag{6.16b}$$

As usual, $S(i, 1, 2)$ and $S(i, 2, 1)$ are transposes of each other. Also submatrix $S(i, 2, 2)$ becomes equal to $S(i, 1, 1)$ when the signs of coefficients along the dotted line in Eq. (6.16a) are reversed, except that of β_i/h_i. This line has no other significance.

It can be verified that the submatrix $[t_i]$ for the element of Fig. 6.3 is

$$[t_i] = \begin{bmatrix} x^g x & x^g y & x^g z & 0 & 0 & 0 \\ y^g x & y^g y & y^g z & 0 & 0 & 0 \\ z^g x & z^g y & z^g z & 0 & 0 & 0 \\ 0 & 0 & 0 & x^g x & x^g y & x^g z \\ 0 & 0 & 0 & y^g x & y^g y & y^g z \\ 0 & 0 & 0 & z^g x & z^g y & z^g z \end{bmatrix} \tag{6.17}$$

in which $x^g x$, for example, denotes the cosine of the angle between the X^g and

the x axes; similarly, $z^g y$ is the cosine of the angle between the Z^g and the y axes, etc.

As in plane frames, $[t_i]$ here also transforms the 'local' extremity forces and displacements to their 'global' counterparts. Thus, the local and global internal forces at extremity a, for example, are related as

$$\begin{Bmatrix} F^g_{ai} \\ Q^g_{ai} \\ H^g_{ai} \\ M^g_{xai} \\ M^g_{yai} \\ M^g_{zai} \end{Bmatrix} = [t_i] \begin{Bmatrix} F_{ai} \\ Q_{ai} \\ H_{ai} \\ M_{xai} \\ M_{yai} \\ M_{zai} \end{Bmatrix} \tag{6.18}$$

Eqs. (6.17) and (6.18) show that the translational and rotational forces are transformed independently of each other. This is obviously also true in the case of displacements. Furthermore, the transformation matrix

$$[T_i] = \begin{bmatrix} t_i & 0 \\ 0 & t_i \end{bmatrix}$$

is now a 12 × 12 matrix; it is also orthogonal and satisfies Eq. (6.2).

To solve a given problem, the $S^g(i, J, K)$'s of the individual elements are first calculated from Eqs. (6.11), (6.16) and (6.17). These are then assembled by the methods of Chapter 3 to form $[K^g]$. The problem will finally be solved in terms of Eq. (6.14) in which the listings of the nodal vectors are as follows:

$$\{P^g\} = \{F^g_1 \; Q^g_1 \; H^g_1 \; M^g_{x1} \; M^g_{y1} \; M^g_{z1} \; \cdots \; F^g_N \; Q^g_N \; H^g_N \; M^g_{xN} \; M^g_{yN} \; M^g_{zN}\}^T \tag{6.19}$$

and

$$\{\delta^g\} = \{u^g_1 \; v^g_1 \; w^g_1 \; \theta^g_{x1} \; \theta^g_{y1} \; \theta^g_{z1} \; \cdots \; u^g_N \; v^g_N \; w^g_N \; \theta^g_{xN} \; \theta^g_{yN} \; \theta^g_{zN}\}^T \tag{6.20}$$

in which N is the total number of nodes in the frame. The nodal forces and displacements have been defined in Fig. 6.3.

6.7 Calculation of internal forces

Once the nodal displacements $\{\delta^g\}$ have been calculated, the internal forces within individual elements will be calculated as follows.

Consider, for example, the plane frame element i whose extremities a and b are attached to the nodes n and k, respectively. Then, in order to be compatible (section 3.1), the displacements of extremities a and b must be equal to those of

n and k respectively; i.e.,

$$\begin{Bmatrix} u^{g}_{ai} \\ v^{g}_{ai} \\ \theta^{g}_{ai} \\ u^{g}_{bi} \\ v^{g}_{bi} \\ \theta^{g}_{bi} \end{Bmatrix} = \begin{Bmatrix} u^{g}_{n} \\ v^{g}_{n} \\ \theta^{g}_{n} \\ u^{g}_{k} \\ v^{g}_{k} \\ \theta^{g}_{k} \end{Bmatrix} \tag{6.21}$$

in which the right hand vector lists the computed global nodal displacements at n and k. It is now a simple matter to show from Eqs. (6.4), (6.5) and (6.21) that the following equation gives the internal forces within i in its *local axes:*

$$\begin{Bmatrix} F_{ai} \\ Q_{ai} \\ M_{ai} \\ F_{bi} \\ Q_{bi} \\ M_{bi} \end{Bmatrix} = [K_i T_i^{\mathsf{T}}] \begin{Bmatrix} u^{g}_{n} \\ v^{g}_{n} \\ \theta^{g}_{n} \\ u^{g}_{k} \\ v^{g}_{k} \\ \theta^{g}_{k} \end{Bmatrix} \tag{6.22}$$

A similar equation is easily derived for space frames.

Uniformly distributed loads

It is easily demonstrated that when element i of a frame structure carries a uniformly distributed load, the 'global equivalent force vector', which we shall denote by $\{P^{g}_{e}\}$, will be given by

$$\{P^{g}_{e}\} = [T_i]\{P_e\} \tag{6.23}$$

Consider, for example, the plane frame element shown in Fig. 6.4.

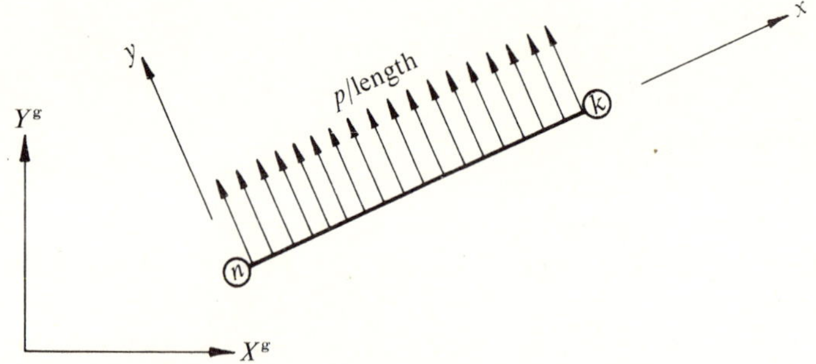

Fig. 6.4 A rigid-jointed plane frame element carrying a uniformly distributed load of intensity p.

$\{P_e\}$ for this element is

$$\{P_e\} = \begin{Bmatrix} F_n \\ Q_n \\ M_n \\ F_k \\ Q_k \\ M_k \end{Bmatrix} = \begin{Bmatrix} 0 \\ ph_i/2 \\ ph_i^2/12 \\ 0 \\ ph_i/2 \\ -ph_i^2/12 \end{Bmatrix} \tag{6.24}$$

Eq. (6.24) is obtained merely by introducing into Eq. (5.8) the 'local' axial forces F_n and F_k; these are both zero here since p does not induce axial forces at the nodes n and k. Then, using $[T_i]$ defined by Eq. (6.1), $\{P_e^g\}$ for the plane frame element can be calculated from Eqs. (6.23) and (6.24).

An expression similar to Eq. (6.24) is easily derived for space frame elements.

6.8 An example

As a typical application of the method developed in this chapter, let us now solve the plane rigid-jointed frame problem of Fig. 6.5a. A subdivision of the frame into finite elements is shown in Fig. 6.5b. The member details are as follows:

Member, i	a_i (m²)	h_i (m)	I_i (m⁴)	α_i (°)
1	0·108	4·5	0·00110	60
2	0·108	4·0	0·00110	0
3	0·090	3·0	0·00061	90

For all members: $E = 20$ GN/m²

Since all displacements vanish at the nodes 1 and 4, these need not be included in the $[K^g]$ matrix of the frame (method 2, section 5.1). Then, following the method of section 3.4 we can show that

$$[K^g] = \begin{matrix} & 2 & 3 \\ 2 & S^g(1,2,2) + S^g(2,1,1) & S^g(2,1,2) \\ 3 & S^g(2,2,1) & S^g(2,2,2) + S^g(3,2,2) \end{matrix} \tag{6.25}$$

The $S^g(i, J, K)$'s will now be calculated from Eqs. (3.4), (6.9) and (6.11). For example, by setting $i = 1$, $J = 2$ and $K = 2$ into Eq. (6.11) we obtain

$$S^g(1, 2, 2) = (t_1 \,.\, S(1, 2, 2) \,.\, t_1^T)$$

or, substituting the property values of element 1 into this equation, and ignoring

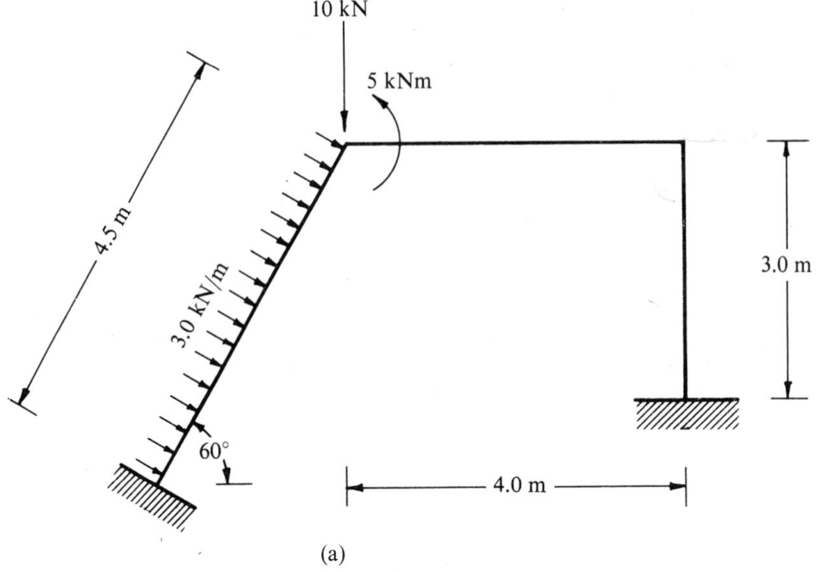

(a)

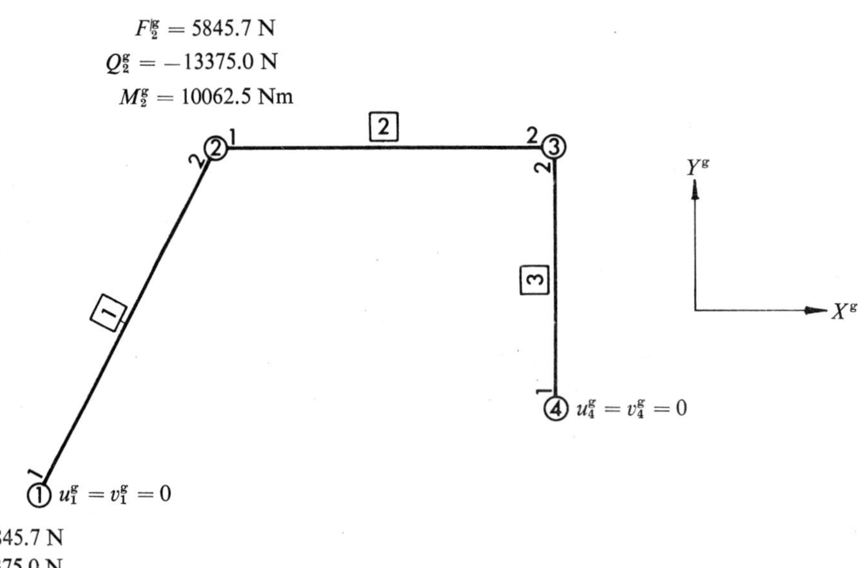

(b)

Fig. 6.5 (a) A rigid-jointed plane frame. (b) A finite element model of (a).

torsion in Eq. (3.4c)

$$S^g(1, 2, 2) = E_1 \begin{bmatrix} \cos 60^\circ & -\sin 60^\circ & 0 \\ \sin 60^\circ & \cos 60^\circ & 0 \\ 0 & 0 & 1 \end{bmatrix}$$

$$\times \begin{bmatrix} a_1/h_1 & 0 & 0 \\ 0 & 12I_1/h_1^3 & -6I_1/h_1^2 \\ 0 & -6I_1/h_1^2 & 4I_1/h_1 \end{bmatrix} \begin{bmatrix} \cos 60^\circ & \sin 60^\circ & 0 \\ -\sin 60^\circ & \cos 60^\circ & 0 \\ 0 & 0 & 1 \end{bmatrix}$$

$$= 20 \times 10^7 \begin{bmatrix} 0{\cdot}61087 & 1{\cdot}03296 & 0{\cdot}02823 \\ 1{\cdot}03296 & 1{\cdot}80362 & -0{\cdot}01630 \\ 0{\cdot}02823 & -0{\cdot}01630 & 0{\cdot}09778 \end{bmatrix}$$

Having calculated all the required $S^g(i, J, K)$'s in this way and on substituting them into Eq. (6.25), we obtain

$$[K^g] = 20 \times 10^7 \begin{bmatrix} 3{\cdot}31087 & & & & & \\ 1{\cdot}03296 & 1{\cdot}82424 & & \text{symmetrical} & & \\ 0{\cdot}02823 & 0{\cdot}02495 & 0{\cdot}20778 & & & \\ -2{\cdot}70000 & 0{\cdot}00000 & 0{\cdot}00000 & 2{\cdot}72667 & & \\ 0{\cdot}00000 & -0{\cdot}02062 & -0{\cdot}04125 & 0{\cdot}00000 & 3{\cdot}02062 & \\ 0{\cdot}00000 & 0{\cdot}04125 & 0{\cdot}05500 & 0{\cdot}04000 & -0{\cdot}04125 & 0{\cdot}19000 \end{bmatrix} \quad (6.26)$$

The uniformly distributed load on element 1 will be treated by Eqs. (6.23) and (6.24); setting $n = 1$, $k = 2$, $h_i = 4{\cdot}5$ and $p = -3000$ into the latter equation, we get

$$\{P_e\} = \begin{Bmatrix} F_1 \\ Q_1 \\ M_1 \\ F_2 \\ Q_2 \\ M_2 \end{Bmatrix} = \begin{Bmatrix} 0 \\ -6750{\cdot}0 \\ -5062{\cdot}5 \\ 0 \\ -6750{\cdot}0 \\ 5062{\cdot}5 \end{Bmatrix}$$

in which all quantities are in the SI units. Then, substitution of this $\{P_e\}$ and $[T_1]$ ($[T_i]$ is defined in Eq. (6.1)) into Eq. (6.23) leads to

$$\{P_e^g\} = \begin{bmatrix} \cos 60^\circ & -\sin 60^\circ & 0 & 0 & 0 & 0 \\ \sin 60^\circ & \cos 60^\circ & 0 & 0 & 0 & 0 \\ 0 & 0 & 1 & 0 & 0 & 0 \\ 0 & 0 & 0 & \cos 60^\circ & -\sin 60^\circ & 0 \\ 0 & 0 & 0 & \sin 60^\circ & \cos 60^\circ & 0 \\ 0 & 0 & 0 & 0 & 0 & 1 \end{bmatrix} \begin{Bmatrix} 0{\cdot}0 \\ -6750{\cdot}0 \\ -5062{\cdot}5 \\ 0{\cdot}0 \\ -6750{\cdot}0 \\ 5062{\cdot}5 \end{Bmatrix} \tag{6.27}$$

or, performing the multiplication,

$$\{P_e^g\} = \begin{Bmatrix} F_1^g \\ Q_1^g \\ M_1^g \\ F_2^g \\ Q_2^g \\ M_2^g \end{Bmatrix} = \begin{Bmatrix} 5845{\cdot}7 \\ -3375{\cdot}0 \\ -5062{\cdot}5 \\ 5845{\cdot}7 \\ -3375{\cdot}0 \\ 5062{\cdot}5 \end{Bmatrix} \tag{6.28}$$

It is now clear from Eq. (6.28) and Fig. 6.5a that the total nodal loads are as follows:

$$F_1^g = 5845{\cdot}7 \text{ N}$$
$$Q_1^g = -3375{\cdot}0 \text{ N}$$
$$M_1^g = -5062{\cdot}5 \text{ Nm}$$
$$F_2^g = 5845{\cdot}7 \text{ N}$$
$$Q_2^g = -3375{\cdot}0 - 10000{\cdot}0 = -13375{\cdot}0 \text{ N}$$

and

$$M_2^g = 5062{\cdot}5 + 5000{\cdot}0 = 10062{\cdot}5 \text{ Nm}$$

However, since nodes 1 and 4 are not included in $[K^g]$, the external force vector here is

$$\{P^g\} = \{F_2^g \quad Q_2^g \quad M_2^g \quad F_3^g \quad Q_3^g \quad M_3^g\}^T$$
$$= \{5845{\cdot}7 \quad -13375{\cdot}0 \quad 10062{\cdot}5 \quad 0{\cdot}0 \quad 0{\cdot}0 \quad 0{\cdot}0\}^T \tag{6.29}$$

The nodal displacements $\{\delta^g\}$ will now be found by making substitutions from Eqs. (6.26) and (6.29) into Eq. (6.14) and solving; a computer solution gave the following displacements:

node k	u_k^g (m)	v_k^g (m)	θ_k^g (rad)
2	0·001278	−0·000760	0·000201
3	0·001268	−0·000005	−0·000161

The internal forces within individual elements will now be found from Eq. (6.22). Consider element 1 for example; setting $n = 1$ and $k = 2$ and noting that all displacements at node 1 are zero, we obtain from this equation

$$\begin{Bmatrix} F_{a1} \\ Q_{a1} \\ M_{a1} \\ F_{b1} \\ Q_{b1} \\ M_{b1} \end{Bmatrix} = [K_1][T_1]^{\mathrm{T}} \begin{Bmatrix} 0 \\ 0 \\ 0 \\ 0{\cdot}001278 \\ -0{\cdot}000760 \\ 0{\cdot}000201 \end{Bmatrix} = \begin{Bmatrix} 8986{\cdot}1 \\ 5620{\cdot}6 \\ 11661{\cdot}0 \\ -8986{\cdot}1 \\ -5620{\cdot}6 \\ 13631{\cdot}6 \end{Bmatrix}$$

in which all quantities are in SI units (note that in Fig. 6.5b the digits 1 and 2 identify the extremities a and b respectively). However, since $\{P_{\mathrm{e}}^{\mathrm{g}}\}$ was used to represent the uniformly distributed load acting on member 1, the above computed internal forces must now be corrected by using Eq. (5.14); thus, the true internal forces in element 1 are

$$\begin{Bmatrix} F_{a1} \\ Q_{a1} \\ M_{a1} \\ F_{b1} \\ Q_{b1} \\ M_{b1} \end{Bmatrix} = \begin{Bmatrix} 8986{\cdot}1 - 0 \\ 5620{\cdot}6 + 6750{\cdot}0 \\ 11661{\cdot}0 + 5062{\cdot}5 \\ -8986{\cdot}1 - 0 \\ -5620{\cdot}6 + 6750{\cdot}0 \\ 13631{\cdot}6 - 5062{\cdot}5 \end{Bmatrix} = \begin{Bmatrix} 8986{\cdot}1 \\ 12370{\cdot}6 \\ 16723{\cdot}5 \\ -8986{\cdot}1 \\ 1129{\cdot}4 \\ 8569{\cdot}1 \end{Bmatrix}$$

The following are the internal forces obtained for the frame of Fig. 6.5 (all quantities are in SI units):

Element i	F_{ai}	Q_{ai}	M_{ai}	F_{bi}	Q_{bi}	M_{bi}
1	8986·1	12370·6	16723·5	−8986·1	1129·4	8569·1
2	5471·2	−2782·6	−3569·1	−5471·2	2782·6	−7561·2
3	2782·6	5471·2	8852·2	−2782·6	−5471·2	7561·2

According to our convention the axial force in element i is tensile when F_{ai} is negative and F_{bi} positive, while it is compressive when these signs are reversed. Thus, the axial (internal) force in element 1 is compressive, its magnitude being 8986·1 N. It may be verified that the finite element solution of this problem coincides with its exact solution.

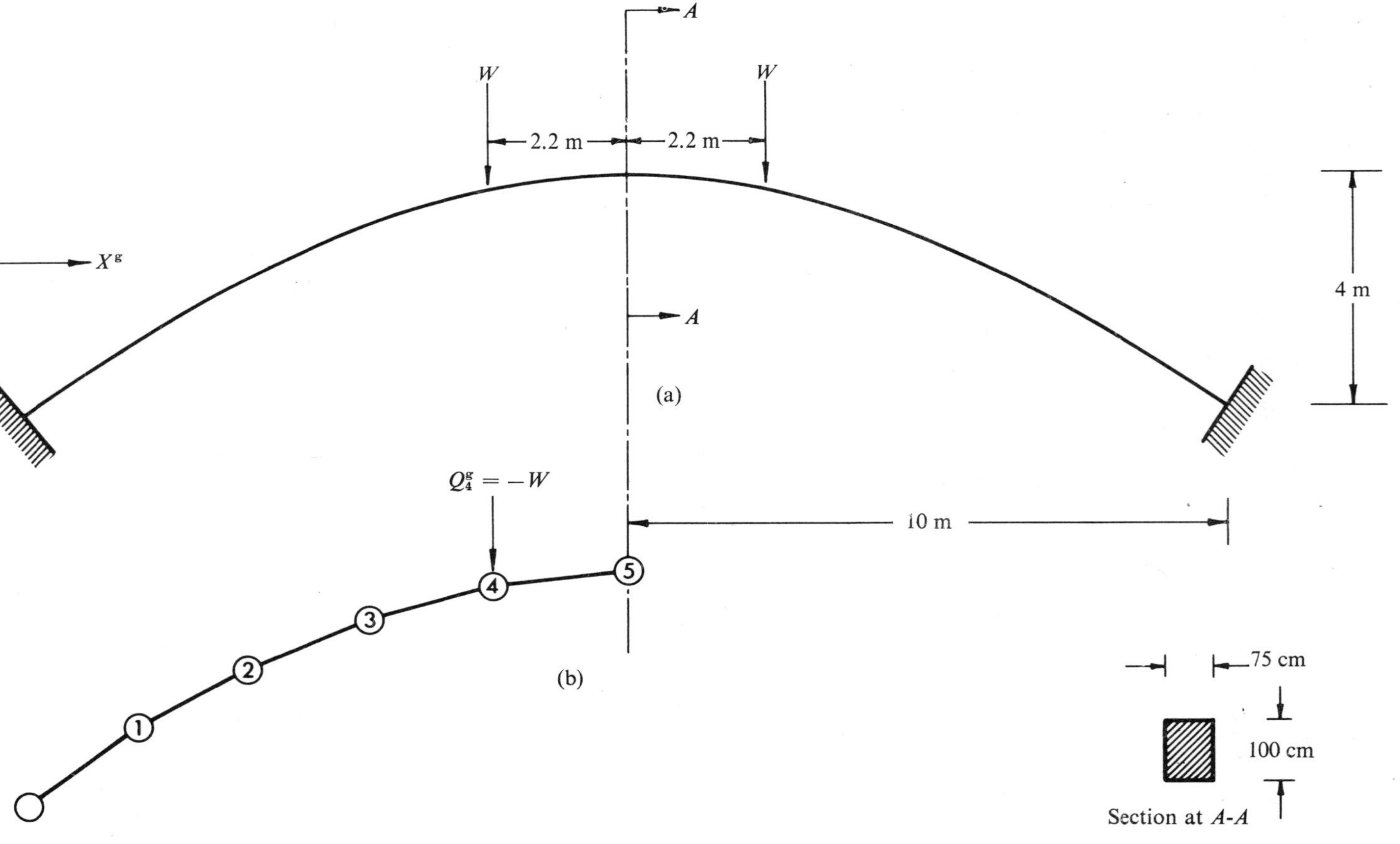

Fig. 6.6 (a) A uniform, symmetric, parabolic plane arch. (b) A proposed finite element model of (a).

6.9 Curved beams

The $S^g(i, J, K)$'s and the transformation matrices developed in this chapter can be used effectively to analyze uniform or non-uniform beams which are curved on a plane or in space, by treating them as assemblages of linear beam elements. The analysis then proceeds exactly as in the case of rigid jointed frames.

As an example, consider the plane symmetrical parabolic arch of Fig. 6.6a which has a uniform cross section. Owing to its symmetry, any symmetrical half of the arch need only be considered; Fig. 6.6b shows a subdivision of a symmetrical half into a number of beam elements of equal length (they need not be equal).

The boundary conditions here are that all displacements vanish at the abutments; the nodes representing these need not therefore by considered. Also, owing to symmetry,

$$u_5^g = \theta_5^g = 0 \tag{6.30}$$

as indicated in section 5.3.

The $S^g(i, J, K)$'s of all the elements are now calculated from Eqs. (3.4), (6.9) and (6.11). These are then assembled in the usual way to give the overall global stiffness matrix $[K^g]$. As the degrees of freedom here is 3 (torsion not considered), $[K^g]$ will have a size of 13×13, since its rows and columns corresponding to the displacements of Eq. (6.30) will be deleted (see method 2, section 5.1). Also, since the only external load is

$$Q_4^g = -W,$$

all the elements of Eq. (6.12) other than Q_4^g will be zero.

A computer solution for this problem gives the following displacements:

Node k	u_k^g(m)*	v_k^g(m)*	θ_k^g(rad)*
1	−11·7741	4·7900	4·6532
2	−14·8981	−3·3455	−11·3185
3	−3·9367	−46·0934	−26·9293
4	4·2056	−103·3248	−18·9289
5		−122·0519	

* Multiplier = W/E; W and E are in SI units.

It is obvious that in this and similar problems the results will improve and converge to their exact values as the number of elements respresenting the curved structure is increased.

7 The pin-jointed frame

7.1 General

A pin-jointed frame is similar in geometry to a rigid-jointed frame, except that its members are interconnected by means of hinges (pin-joints). These frames are designed to sustain external point loads, usually applied at the hinges, and no external force is usually applied on the members. Since an ideal hinge is frictionless and therefore unable to sustain moments, the members are subjected to internal axial forces only. Furthermore, as the magnitude of the internal force occurring within a given member remains constant throughout it, each member of the frame can be represented by a single finite element. Thus, in the finite element model of a pin-jointed frame each element represents a member and each hinge a node. Also, owing to the invariance of internal forces within individual members, a finite element solution of the frame will coincide with its exact solution, arrived at *via* conventional methods.

The symbolism of Chapter 6 will also be used in this chapter to identify quantities which refer to the 'local' and 'global' coordinate systems. Since each member is subjected to an internal axial force only, its 'element stiffness matrix', $[K_i]$, can be obtained from Eq. (2.8) merely by deleting from this equation all the rows and columns which refer to forces other than the axial forces. Thus, the counterpart of Eq. (2.8) for the frame element (member) i becomes

$$\begin{Bmatrix} F_{ai} \\ F_{bi} \end{Bmatrix} = [K_i] \begin{Bmatrix} u_{ai} \\ u_{bi} \end{Bmatrix} \tag{7.1}$$

where

$$[K_i] = k_i \begin{bmatrix} 1 & -1 \\ -1 & 1 \end{bmatrix}^{*} \tag{7.2}$$

and $k_i = a_iE_i/h_i$. Eqs. (7.1) and (7.2) refer to the 'local' axes of element i.

7.2 Treatment of plane frames

Fig. 7.1 shows the 'local' force and displacement at extremity a of element i, resolved along the global axes. It is clear from this diagram that

$$F^{g}_{ai} = F_{ai}l_i$$

and

$$Q^{g}_{ai} = F_{ai}m_i$$

* Clearly, in this case $S(i, 1, 1) = S(i, 2, 2) = -S(i, 1, 2) = -S(i, 2, 1) = k_i$.

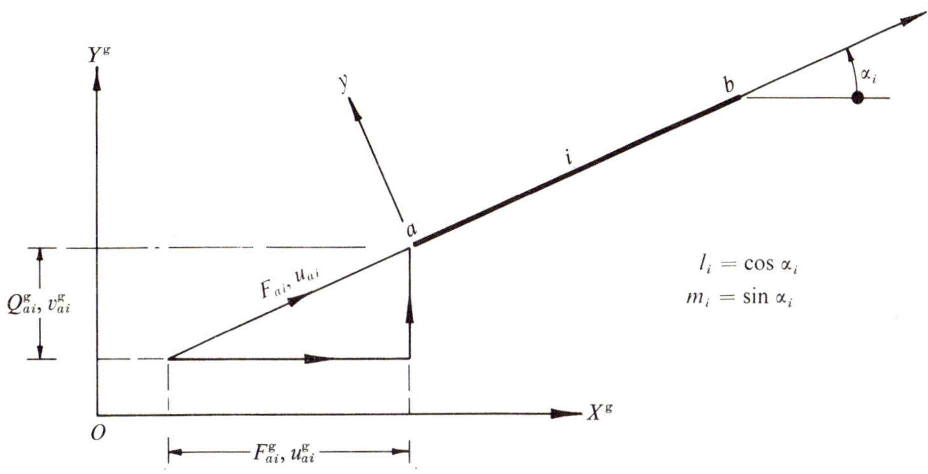

Fig. 7.1 A plane pin-jointed frame element.

For extremity b, we similarly have

$$F^g_{bi} = F_{bi} l_i$$

and

$$Q^g_{bi} = F_{bi} m_i$$

Here, as in Chapter 6, symbols with the superscript g denote items referred to the global axes while, those with subscripts only represent items referred to the local axes of the element. Writing the above four equations in the matrix form,

$$\begin{Bmatrix} F^g_{ai} \\ Q^g_{ai} \\ F^g_{bi} \\ Q^g_{bi} \end{Bmatrix} = \begin{bmatrix} l_i & 0 \\ m_i & 0 \\ 0 & l_i \\ 0 & m_i \end{bmatrix} \begin{Bmatrix} F_{ai} \\ F_{bi} \end{Bmatrix} \tag{7.3}$$

The rectangular matrix in Eq. (7.3) obviously relates the global forces to the local forces. Therefore, by definition (section 6.2),

$$[T_i] = \begin{bmatrix} l_i & 0 \\ m_i & 0 \\ 0 & l_i \\ 0 & m_i \end{bmatrix} \tag{7.4}$$

is the transformation matrix in this case. It is also obvious from Fig. 7.1 that the local displacements will be related to the global displacements through $[T_i]$

as

$$\begin{Bmatrix} u_{ai}^g \\ v_{ai}^g \\ u_{bi}^g \\ v_{bi}^g \end{Bmatrix} = [T_i] \begin{Bmatrix} u_{ai} \\ u_{bi} \end{Bmatrix} \tag{7.5}$$

A multiplication will show that since $l_i^2 + m_i^2 = 1$,

$$[T_i^\mathrm{T} T_i] = [I] \tag{7.6}$$

in which $[I]$ is a 2×2 unit matrix.

Premultiplying both sides of Eq. (7.1) by $[T_i]$, we have

$$[T_i] \begin{Bmatrix} F_{ai} \\ F_{bi} \end{Bmatrix} = [T_i K_i] \begin{Bmatrix} u_{ai} \\ u_{bi} \end{Bmatrix} \tag{7.7}$$

The left hand side of Eq. (7.7) is obviously the same as the right hand side of Eq. (7.3); therefore,

$$\begin{Bmatrix} F_{ai}^g \\ Q_{ai}^g \\ F_{bi}^g \\ Q_{bi}^g \end{Bmatrix} = [T_i K_i] \begin{Bmatrix} u_{ai} \\ u_{bi} \end{Bmatrix} \tag{7.8}$$

Further, premultiplying both sides of Eq. (7.5) by $[T_i^\mathrm{T}]$ and using Eq. (7.6) in the result,

$$\begin{Bmatrix} u_{ai} \\ u_{bi} \end{Bmatrix} = [T_i^\mathrm{T}] \begin{Bmatrix} u_{ai}^g \\ v_{ai}^g \\ u_{bi}^g \\ v_{bi}^g \end{Bmatrix} \tag{7.9}$$

Finally, elimination of $\begin{Bmatrix} u_{ai} \\ u_{bi} \end{Bmatrix}$ between Eqs. (7.8) and (7.9) leads to

$$\begin{Bmatrix} F_{ai}^g \\ Q_{ai}^g \\ F_{bi}^g \\ Q_{bi}^g \end{Bmatrix} = [T_i K_i T_i^\mathrm{T}] \begin{Bmatrix} u_{ai}^g \\ v_{ai}^g \\ u_{bi}^g \\ v_{bi}^g \end{Bmatrix} \tag{7.10}$$

In Eq. (7.10), as in Eq. (6.7), the matrix

$$[K_i^g] = [T_i K_i T_i^\mathrm{T}] \tag{7.11}$$

relates the global extremity forces to the global extremity displacements; $[K_i^g]$ is therefore the 'global' element stiffness matrix of i.

From Eqs. (7.2), (7.4) and (7.11), we can show that

$$[K_i^g] = k_i \begin{bmatrix} l_i & 0 \\ m_i & 0 \\ 0 & l_i \\ 0 & m_i \end{bmatrix} \begin{bmatrix} 1 & -1 \\ -1 & 1 \end{bmatrix} \begin{bmatrix} l_i & m_i & 0 & 0 \\ 0 & 0 & l_i & m_i \end{bmatrix}$$

$$= k_i \begin{bmatrix} l_i^2 & l_im_i & -l_i^2 & -l_im_i \\ l_im_i & m_i^2 & -l_im_i & -m_i^2 \\ -l_i^2 & -l_im_i & l_i^2 & l_im_i \\ -l_im_i & -m_i^2 & l_im_i & m_i^2 \end{bmatrix} \tag{7.12}$$

$[K_i^g]$ is a symmetric matrix, as we would have expected from the reciprocal theorem. Following the procedure of Chapter 3, $[K_i^g]$ will now be split up into its submatrices, $S^g(i, J, K)$'s; by their definition, we can simply write them down by inspection from Eq. (7.12) as

$$S^g(i, 1, 1) = S^g(i, 2, 2) = -S^g(i, 1, 2) = -S^g(i, 2, 1)$$

$$= k_i \begin{bmatrix} l_i^2 & l_im_i \\ l_im_i & m_i^2 \end{bmatrix} \tag{7.13}$$

The global overall stiffness matrix, $[K^g]$, for the whole frame will now be formed by assembling the $S^g(i, J, K)$'s of individual elements by the methods of Chapter 3. The problem then reduces to the solution of Eq. (6.14) with the prescribed boundary conditions. The listings of $\{P^g\}$ and $\{\delta^g\}$ here are as follows:

$$\{P^g\} = \{F_1^gQ_1^g\ F_2^gQ_2^g \cdots F_N^gQ_N^g\}^T \tag{7.14}$$

and

$$\{\delta^g\} = \{u_1^gv_1^g\ u_2^gv_2^g \cdots u_N^gv_N^g\}^T \tag{7.15}$$

where N is the total number of nodes (hinges) in the frame.

7.3 Treatment of space frames

Figure 7.2 shows the 'local' force and displacement at extremity a of a space frame element, resolved along the global axes. Clearly, in this case

$$F_{ai}^g = F_{ai}l_i$$

$$Q_{ai}^g = F_{ai}m_i$$

and

$$H_{ai}^g = F_{ai}n_i$$

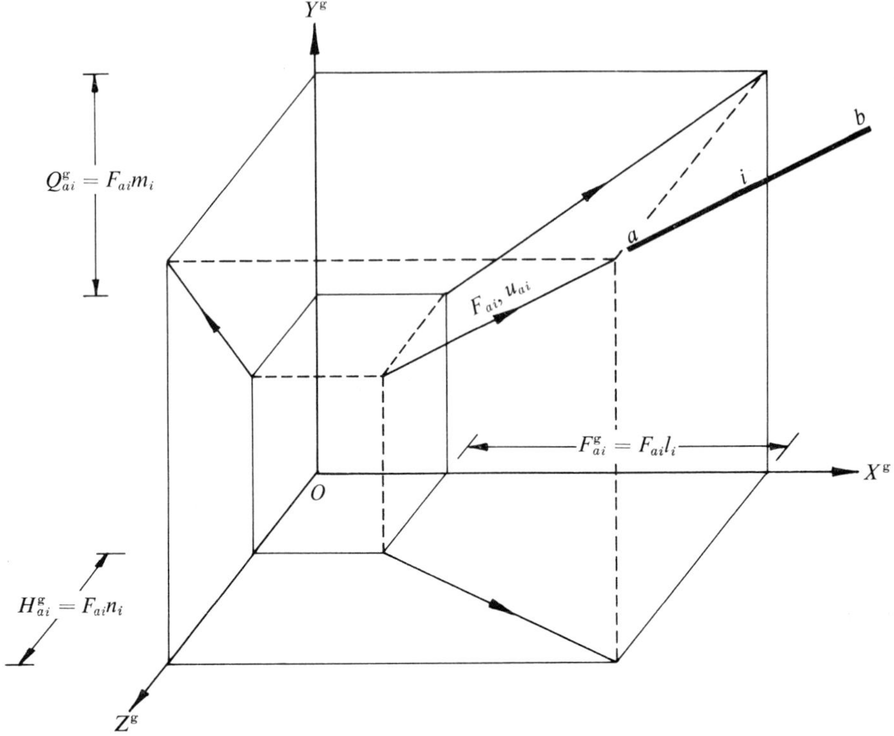

Fig. 7.2 A pin-jointed space frame element. (l_i, m_i and n_i are the direction cosines of i with respect to the global axes).

Similarly for extremity b. We can therefore demonstrate, as in section 7.2, that

$$[T_i] = \begin{bmatrix} l_i & 0 \\ m_i & 0 \\ n_i & 0 \\ 0 & l_i \\ 0 & m_i \\ 0 & n_i \end{bmatrix} \tag{7.16}$$

Since $l_i^2 + m_i^2 + n_i^2 = 1$, $[T_i]$ in Eq. (7.16) also satisfies Eq. (7.6). It is now a simple matter to show from Eqs. (7.2), (7.11) and (7.16) that for a space frame

element

$$S^g(i,1,1) = S^g(i,2,2) = -S^g(i,1,2) = -S^g(i,2,1)$$

$$= k_i \begin{bmatrix} l_i^2 & l_i m_i & l_i n_i \\ m_i l_i & m_i^2 & m_i n_i \\ n_i l_i & n_i m_i & n_i^2 \end{bmatrix} \tag{7.17}$$

The remaining steps leading to the final solution of a space frame problem are identical to those for a plane frame problem, except that the listings of $\{P^g\}$ and $\{\delta^g\}$ are now as follows:

$$\{P^g\} = \{F_1^g Q_1^g H_1^g \; F_2^g Q_2^g H_2^g \cdots F_N^g Q_N^g H_N^g\}^T \tag{7.18}$$

and

$$\{\delta^g\} = \{u_1^g v_1^g w_1^g \; u_2^g v_2^g w_2^g \cdots u_N^g v_N^g w_N^g\}^T \tag{7.19}$$

in which, H_k^g and w_k^g are externally applied force and displacement along Z^g at node k (= 1, 2, . . . N).

7.4 Internal forces in members

The internal forces within individual frame members can be determined from the computed nodal displacements as follows:

Consider the plane frame member i whose extremities a and b are attached to the nodes n and k respectively. Eliminating the vector $\begin{Bmatrix} u_{ai} \\ u_{bi} \end{Bmatrix}$ between Eqs. (7.1) and (7.9), we have

$$\begin{Bmatrix} F_{ai} \\ F_{bi} \end{Bmatrix} = [K_i T_i^T] \begin{Bmatrix} u_{ai}^g \\ v_{ai}^g \\ u_{bi}^g \\ v_{bi}^g \end{Bmatrix} \tag{7.20}$$

Now, in order to be compatible, the displacements of extremities a and b must be equal to those of the nodes n and k respectively. We can therefore write Eq. (7.20) also as

$$\begin{Bmatrix} F_{ai} \\ F_{bi} \end{Bmatrix} = [K_i T_i^T] \begin{Bmatrix} u_n^g \\ v_n^g \\ u_k^g \\ v_k^g \end{Bmatrix} \tag{7.21}$$

or, substituting for $[K_i]$ and $[T_i]$ from Eqs. (7.2) and (7.4) respectively, and

multiplying

$$\begin{Bmatrix} F_{ai} \\ F_{bi} \end{Bmatrix} = k_i \begin{bmatrix} l_i & m_i & -l_i & -m_i \\ -l_i & -m_i & l_i & m_i \end{bmatrix} \begin{Bmatrix} u_n^g \\ v_n^g \\ u_k^g \\ v_k^g \end{Bmatrix} \tag{7.22}$$

Thus, when the nodal displacements u_n^g, v_n^g etc. have been calculated, the internal force F_{ai} (or F_{bi}) within element i will be found from Eq. (7.22). It is clear from Eq. (7.22) that F_{ai} and F_{bi} are equal in magnitude but opposite in sense. According to our convention, if F_{ai} is negative and F_{bi} positive, as in Fig. 7.3, then element i will be in tension, while it will be in compression when these signs are reversed.

For the space frame member i it is easy to demonstrate in this way that

$$\begin{Bmatrix} F_{ai} \\ F_{bi} \end{Bmatrix} = k_i \begin{bmatrix} l_i & m_i & n_i & -l_i & -m_i & -n_i \\ -l_i & -m_i & -n_i & l_i & m_i & n_i \end{bmatrix} \begin{Bmatrix} u_n^g \\ v_n^g \\ w_n^g \\ u_k^g \\ v_k^g \\ w_k^g \end{Bmatrix} \tag{7.23}$$

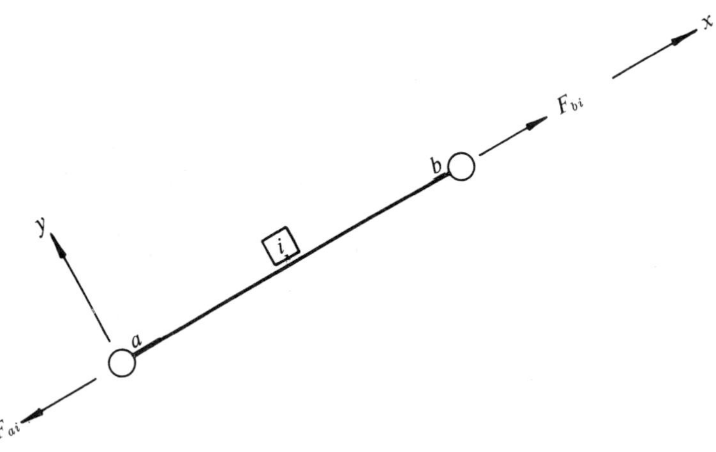

Fig. 7.3 A frame member in tension.

7.5 An example

Let us now solve the simple plane frame problem of Fig. 7.4a by the method developed in this chapter. It is given that all members of this frame have the same uniform cross sectional area a and Young's modulus E. The finite element model of the frame is shown in Fig. 7.4b. Clearly, the member details are as follows:

Member, i	α_i (°)	Length (h_i)
1	30	$2H$
2	45	$H\sqrt{2}$
3	60	$2H/\sqrt{3}$
4	90	H

Following the method of section 3.4, the overall stiffness matrix $[K^g]$ of the frame is now assembled. Since all displacements vanish at the nodes 1, 2, 3 and 4, the boundary conditions of the frame are

$$u_1^g = v_1^g = u_2^g = v_2^g = u_3^g = v_3^g = u_4^g = v_4^g = 0 \tag{7.24}$$

Consequently, using method 2 of section 5.1 the rows and columns of $[K^g]$ which refer to these displacements will not be taken into consideration. It is easily verified that for this problem Eq. (6.14) thus becomes

$$[S^g(1, 2, 2) + S^g(2, 2, 2) + S^g(3, 2, 2) + S^g(4, 2, 2)]\begin{Bmatrix} u_5^g \\ v_5^g \end{Bmatrix} = \begin{Bmatrix} 0 \\ W \end{Bmatrix} \tag{7.25}$$

The right hand vector of Eq. (7.25) derives from the fact that the only external force on the frame is $Q_5^g = W$.

We can now show from Eq. (7.13) that

$$S^g(1, 2, 2) = r\begin{bmatrix} 0{\cdot}37500 & 0{\cdot}21651 \\ 0{\cdot}21651 & 0{\cdot}12500 \end{bmatrix}$$

$$S^g(2, 2, 2) = r\begin{bmatrix} 0{\cdot}35355 & 0{\cdot}35355 \\ 0{\cdot}35355 & 0{\cdot}35355 \end{bmatrix}$$

$$S^g(3, 2, 2) = r\begin{bmatrix} 0{\cdot}21651 & 0{\cdot}37500 \\ 0{\cdot}37500 & 0{\cdot}64952 \end{bmatrix}$$

and

$$S^g(4, 2, 2) = r\begin{bmatrix} 0{\cdot}00000 & 0{\cdot}00000 \\ 0{\cdot}00000 & 1{\cdot}00000 \end{bmatrix}$$

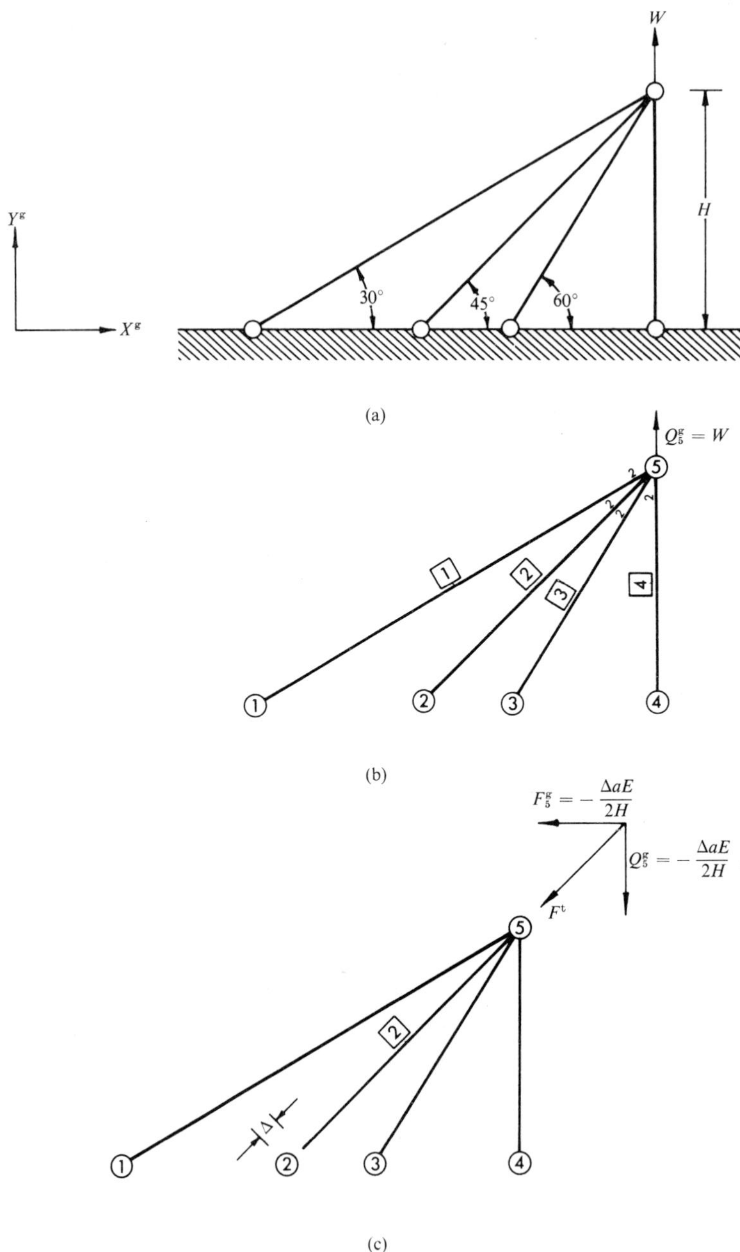

Fig. 7.4 (a) A plane pin-jointed frame; all members have the same area of cross section and Young's modulus. (b) Finite element model of (a). (c) Lack of fit, when member 2 is too short by the amount Δ.

in which $r = aE/H$. Substitution of these $S^g(i, J, K)$'s into Eq. (7.25) leads to

$$r\begin{bmatrix} 0{\cdot}94506 & 0{\cdot}94506 \\ 0{\cdot}94506 & 2{\cdot}12807 \end{bmatrix}\begin{Bmatrix} u_5^g \\ v_5^g \end{Bmatrix} = \begin{Bmatrix} 0 \\ W \end{Bmatrix} \tag{7.26}$$

Solving Eq. (7.26), we get

$$u_5^g = -0{\cdot}84523HW/aE$$

and

$$v_5^g = 0{\cdot}84523HW/aE$$

Having obtained the nodal displacements, the internal member forces can be calculated from Eq. (7.22). Setting $n = 1$ and $k = 5$ in this equation, for example, the internal force in member 1 is found to be

$$\begin{Bmatrix} F_{a1} \\ F_{b1} \end{Bmatrix} = \frac{aE}{2H}\begin{bmatrix} 0{\cdot}86603 & 0{\cdot}50000 & -0{\cdot}86603 & -0{\cdot}50000 \\ -0{\cdot}86603 & -0{\cdot}50000 & 0{\cdot}86603 & 0{\cdot}50000 \end{bmatrix}\begin{Bmatrix} u_1^g \\ v_1^g \\ u_5^g \\ v_5^g \end{Bmatrix}$$

$$= \begin{Bmatrix} 0{\cdot}1547W \\ -0{\cdot}1547W \end{Bmatrix}$$

Clearly, F_{a1} is positive and F_{b1} negative here. Therefore, according to our convention (section 7.4) the force in member 1 is compressive, its magnitude being $0{\cdot}1547W$. Similarly, we can demonstrate that the remaining member forces of the frame are

Member	Magnitude	Type
2	0·0000	
3	0·2679*W*	Tensile
4	0·8452*W*	Tensile

Lack of fit problems

Suppose that in the frame of Fig. 7.4a the force W is removed and that member 2 (Fig. 7.4c) is initially too short by the amount Δ. Now, if we force this member into position between the hinges (nodes) 2 and 5, then this will obviously induce internal forces into the frame members even though the frame is free from externally applied forces. Redundant frame problems of this type in which a member(s) is initially either too short or too long compared to its exact length are generally referred to as 'lack of fit' problems.

In the finite element method such problems are dealt with by treating the lack of fit forces as external forces applied at the appropriate nodes. Consider

the typical problem of Fig. 7.4c; since member 2 is initially too short, it is easy to see that the magnitude of the lack of fit force here is

$$F^t = \Delta a E/\sqrt{2}H$$

and that it is a tensile force in member 2. Consequently, we shall treat this force as an external force acting at node 5 as shown in Fig. 7.4c (note that an equal force acting in the opposite direction must also be applied at node 2; we need not consider this force, however, since all displacements vanish at this node). The global components of F^t are obviously

$$F_5^g = -\Delta a E/2H$$

and

$$Q_5^g = -\Delta a E/2H$$

Consequently, Eq. (7.26) now becomes

$$r\begin{bmatrix} 0{\cdot}94506 & 0{\cdot}94506 \\ 0{\cdot}94506 & 2{\cdot}12807 \end{bmatrix}\begin{Bmatrix} u_5^g \\ v_5^g \end{Bmatrix} = -\frac{\Delta a E}{2H}\begin{Bmatrix} 1 \\ 1 \end{Bmatrix}$$

Solving this equation we have

$$u_5^g = -0{\cdot}5291\Delta$$

and

$$v_5^g = 0{\cdot}0000$$

The individual member forces can now be calculated as before.

8 Fundamentals of linear elasticity

The derivation of the stiffness matrix of the beam element in Chapter 2 was deliberately restricted within the confines of elementary structural mechanics with which the beginner is familiar. In the next chapter we shall present a general procedure, based on the principle of virtual work, from which the stiffness and other characteristics of any finite element can be derived. As the understanding and application of this procedure is facilitated by some knowledge of elasticity, we shall now discuss briefly those aspects of linear elasticity which are of interest in the general context of finite elements.

8.1 Displacements and strains

When an elastic body, whose rigid-body displacements have been suppressed and which satisfies assumptions 1 and 2 of section 2.1, is acted upon by one or more forces, it deforms and every particle within it is displaced to a new position. Let A and A' in Fig. 8.1 be the undeformed and deformed positions respectively of a typical particle within the body; then, the total displacement at A is clearly AA' whose Cartesian components are u, v and w along the x, y and z axes, respectively. As displacements in general vary from point to point, u, v and w are all functions of position.

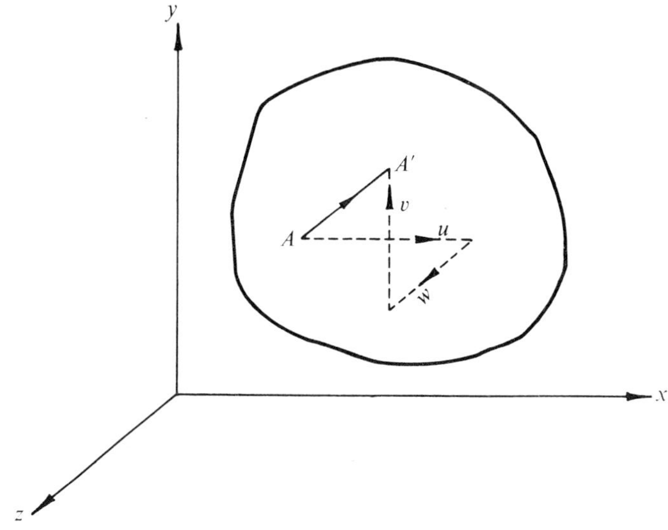

Fig. 8.1 Displacement components in three dimensions.

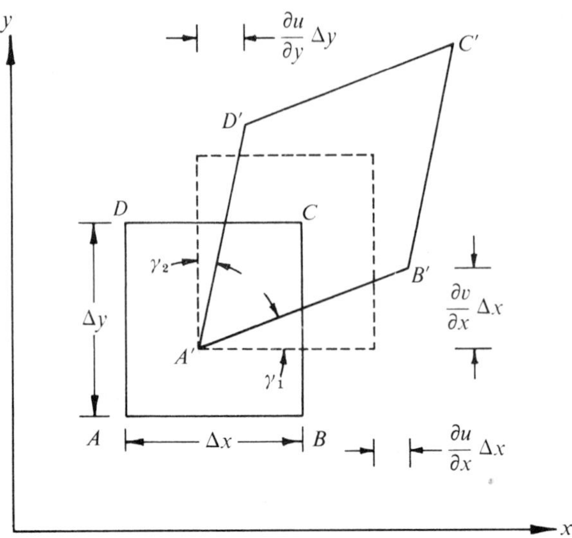

Fig. 8.2 Deformations of a plane element.

The total deformation of the body can be expressed as a combination of 'normal' and 'shear' strains within it. Consider the differential element $ABCD$, shown in Fig. 8.2, which has been isolated from a two dimensional undeformed body, and let $A'B'C'D'$ be its deformed shape when the body is subjected to the action of a system of forces. It is clear from this diagram that the increase in the projection of side AB on the x axis due to deformation is equal to the amount

$$\frac{\partial u}{\partial x}\,\Delta x$$

The strain e_{xx}, which is termed as the 'normal' strain along x, is by definition the ratio of the above amount to the undeformed projection of AB on the x axis, which is Δx; that is $e_{xx} = \partial u/\partial x$. [*Notation* The first suffix denotes the direction and the second identifies the plane. Thus, e_{xy} for example is the strain component along x, and lies on a plane which is normal to the y axis. These suffixes also identify the stresses.] The remaining normal strains e_{yy} and e_{zz}, which are along the y and z axes respectively, can also be expressed in terms of the respective displacements in this way; it is easily verified that in three dimensions

$$e_{xx} = \frac{\partial u}{\partial x}, \qquad e_{yy} = \frac{\partial v}{\partial y} \quad \text{and} \quad e_{zz} = \frac{\partial w}{\partial z} \tag{8.1}$$

The normal strains are associated with changes in the volume of the body without any change in its shape. Indeed, it can be demonstrated that the sum $(e_{xx} + e_{yy} + e_{zz})$, which is called the 'cubical dilatation' or simply 'dilatation', represents the volumetric strain of the body.

The 'shear' strains, on the other hand, represent angular distortions of the body without any change in its volume. Thus, in the case of Fig. 8.2, the shear strain e_{xy} will be defined as

$$e_{xy} = \gamma_1 + \gamma_2 \tag{8.2}$$

where γ_1 and γ_2 are the angles of distortion of the element $ABCD$. It is easily demonstrated that

$$\gamma_1 \approx \tan \gamma_1 = \frac{\dfrac{\partial v}{\partial x}}{1 + \dfrac{\partial u}{\partial x}}$$

As we are dealing with very small deformations, the term $\partial u/\partial x$ ($= e_{xx}$) in the denominator above is negligible compared with unity. We can therefore write

$$\gamma_1 = \frac{\partial v}{\partial x}$$

and similarly

$$\gamma_2 = \frac{\partial u}{\partial y}$$

Then, from Eq. (8.2) it follows that

$$e_{xy} = \frac{\partial u}{\partial y} + \frac{\partial v}{\partial x} \tag{8.3a}$$

It is easily verified in this way that the remaining shear strains e_{yz} and e_{zx} are

$$e_{yz} = \frac{\partial v}{\partial z} + \frac{\partial w}{\partial y} \tag{8.3b}$$

and

$$e_{zx} = \frac{\partial w}{\partial x} + \frac{\partial u}{\partial z} \tag{8.3c}$$

Sometimes the shear strains are also expressed by multiplying the right hand sides of Eqs. (8.3) by $\frac{1}{2}$; this is particularly useful if tensors are used for analysis.

We shall express these six components of elastic strain in the matrix form as

$$\{\epsilon\} = \begin{Bmatrix} e_{xx} \\ e_{yy} \\ e_{zz} \\ e_{xy} \\ e_{yz} \\ e_{zx} \end{Bmatrix} \tag{8.4}$$

The 'elastic' strain vector $\{\epsilon\}$ of Eq. (8.4) obviously refers to a three dimensional body. In plane problems (sections 8.5 and 8.6), on the other hand, if deformations

occur exclusively on the $x - y$ plane, for example, then

$$\{\epsilon\} = \begin{Bmatrix} e_{xx} \\ e_{yy} \\ e_{xy} \end{Bmatrix} \tag{8.4a}$$

8.2 Thermal strains

We have assumed so far that the strains within the body were due solely to the applied forces and occurred under isothermal (constant temperature) conditions. Nearly all materials, however, expand when heated and contract when cooled. Therefore, if the body is exposed to a uniform temperature distribution and allowed to expand or contract freely without any restraint, then only thermal strains will be produced within it. If, on the other hand, its expansion or contraction is restrained, then it will be subjected to thermal stresses. These stresses also arise when the body is exposed to a non-uniform temperature distribution; different parts of the body now expand or contract by varying amounts, depending on the local temperature gradients, and thermal stresses are set up as a result.

For a given set of external forces acting on the body, if we raise or lower its temperature, then thermal strains $\{e_t\}$ will occur in addition to the elastic strains $\{\epsilon\}$. The total strain within the body can then be expressed as the algebraic sum of $\{e_t\}$ and $\{\epsilon\}$. If the body is 'thermally isotropic', meaning that its thermal properties such as coefficients of expansion etc. are the same in all directions, then the thermal strains will result in a change of volume without any distortion. In other words, a thermally isotropic body is unable to sustain shearing thermal strains, and is therefore subject only to normal (dilatational) thermal strains. We shall assume the material of the body always to be thermally isotropic and homogeneous.

The normal thermal strains are easily calculated by noting that when subjected to a temperature change of T, a small linear dimension h of the body changes by the amount αTh where α is its coefficient of linear expansion; clearly, the normal thermal strain here is αT. Applying this to the three coordinate axes in turn, and recalling that all shearing thermal strains are zero owing to thermal isotropy, we have

$$\{e_t\} = \begin{Bmatrix} e_{xx} \\ e_{yy} \\ e_{zz} \\ e_{xy} \\ e_{yz} \\ e_{zx} \end{Bmatrix}_{\text{thermal}} = \alpha T \begin{Bmatrix} 1 \\ 1 \\ 1 \\ 0 \\ 0 \\ 0 \end{Bmatrix} \tag{8.5}$$

in which T is positive, in the case of a rise in temperature.

8.3 Initial strain

'Initial strains' are defined as strains present in the body prior to the application of external forces, including thermal forces. The 'lack of fit' in frames (section 7.5) is a simple example of initial strain.

In the absence of thermal strains, when a body with initial strains is acted upon by external forces, the stresses within it will clearly be caused by the strain

$$\{e\} = \{\epsilon\} - \{e_i\}$$

where the vector $\{e_i\}$ lists all the components of initial strain. In three dimensions the general expression for $\{e_i\}$ is

$$\{e_i\} = \{e_{ixx} \quad e_{iyy} \quad e_{izz} \quad e_{ixy} \quad e_{iyz} \quad e_{izx}\}^T \tag{8.6}$$

in which e_{ixx}, e_{iyy} etc. are the Cartesian components of initial strain. When both thermal and initial strains are present, the stresses within the body will be caused by the strain

$$\{e\} = \{\epsilon\} - \{e_t\} - \{e_i\} \tag{8.7}$$

8.4 Stress-strain relationships

Unless otherwise stated, we shall assume that the material of the body is linearly elastic, isotropic and homogeneous, so that its elastic properties are completely specified by the mutually independent constants E and ν, denoting Young's modulus and Poisson's ratio respectively. Also, assuming for the time being that initial and thermal strains are absent, consider the parallelepiped of Fig. 8.3, each side of which is equal to unity in the unstressed state. According

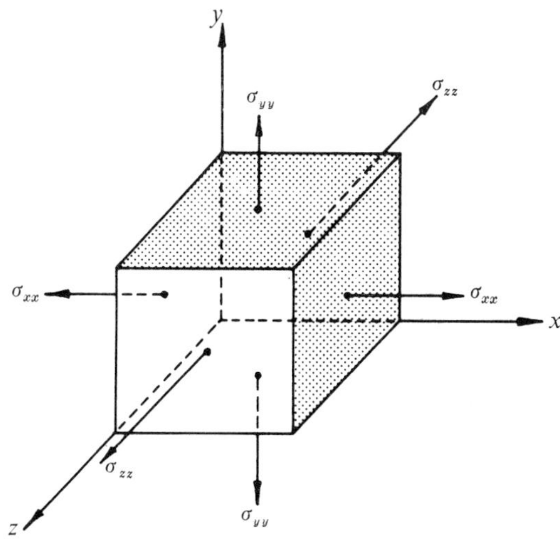

Fig. 8.3 Stresses applied to a parallelepiped of unit sides.

to Hooke's law, the elongation of the parallelepiped along x, which is also its normal strain e_{xx}, is clearly

$$e_{xx} = \frac{1}{E}(\sigma_{xx} - \nu\sigma_{yy} - \nu\sigma_{zz}) \tag{8.8a}$$

in which σ_{jj} ($j = x, y, z$) is the normal stress along j. Similarly,

$$e_{yy} = \frac{1}{E}(\sigma_{yy} - \nu\sigma_{zz} - \nu\sigma_{xx}) \tag{8.8b}$$

and

$$e_{zz} = \frac{1}{E}(\sigma_{zz} - \nu\sigma_{xx} - \nu\sigma_{yy}) \tag{8.8c}$$

Unlike the normal strains, the shear strains are mutually independent, and related to their respective stresses through

$$e_{xy} = \frac{1}{G}\tau_{xy} \tag{8.9a}$$

$$e_{yz} = \frac{1}{G}\tau_{yz} \tag{8.9b}$$

and

$$e_{zx} = \frac{1}{G}\tau_{zx} \tag{8.9c}$$

where τ_{xy}, τ_{yz} and τ_{zx} are shearing stresses, and G the modulus of rigidity of the material of the body, given by

$$G = \frac{E}{2(1 + \nu)} \tag{8.10}$$

The expressions for the above six strain components can now be written in matrix form as

$$\begin{Bmatrix} e_{xx} \\ e_{yy} \\ e_{zz} \\ e_{xy} \\ e_{yz} \\ e_{zx} \end{Bmatrix} = \frac{1}{E}\begin{bmatrix} 1 & -\nu & -\nu & 0 & 0 & 0 \\ -\nu & 1 & -\nu & 0 & 0 & 0 \\ -\nu & -\nu & 1 & 0 & 0 & 0 \\ 0 & 0 & 0 & 2(1+\nu) & 0 & 0 \\ 0 & 0 & 0 & 0 & 2(1+\nu) & 0 \\ 0 & 0 & 0 & 0 & 0 & 2(1+\nu) \end{bmatrix}\begin{Bmatrix} \sigma_{xx} \\ \sigma_{yy} \\ \sigma_{zz} \\ \tau_{xy} \\ \tau_{yz} \\ \tau_{zx} \end{Bmatrix}$$

By inversion, we can write this equation in the following alternative form:

$$\begin{Bmatrix} \sigma_{xx} \\ \sigma_{yy} \\ \sigma_{zz} \\ \tau_{xy} \\ \tau_{yz} \\ \tau_{zx} \end{Bmatrix} = \begin{bmatrix} \lambda + 2G & \lambda & \lambda & 0 & 0 & 0 \\ \lambda & \lambda + 2G & \lambda & 0 & 0 & 0 \\ \lambda & \lambda & \lambda + 2G & 0 & 0 & 0 \\ 0 & 0 & 0 & G & 0 & 0 \\ 0 & 0 & 0 & 0 & G & 0 \\ 0 & 0 & 0 & 0 & 0 & G \end{bmatrix} \begin{Bmatrix} e_{xx} \\ e_{yy} \\ e_{zz} \\ e_{xy} \\ e_{yz} \\ e_{zx} \end{Bmatrix} \tag{8.11}$$

in which λ is the well-known 'Lamé constant', given by

$$\lambda = \frac{\nu E}{(1 + \nu)(1 - 2\nu)}$$

Eq. (8.11) states Hooke's law in three dimensions; we shall write it in the compact symbolic form as

$$\{\sigma\} = [d]\{\epsilon\} \tag{8.12}$$

in which $[d]$ is the 'elasticity matrix', represented by the two-dimensional array of Eq. (8.11); $\{\epsilon\}$ is given by Eq. (8.4) and

$$\{\sigma\} = \{\sigma_{xx} \quad \sigma_{yy} \quad \sigma_{zz} \quad \tau_{xy} \quad \tau_{yz} \quad \tau_{zx}\}^{\mathrm{T}} \tag{8.13}$$

When thermal and initial strains are also present, we must replace $\{\epsilon\}$ by $\{e\}$ of Eq. (8.7); Eq. (8.12) then becomes

$$\{\sigma\} = [d]\{\epsilon\} - [d]\{e_{\mathrm{t}}\} - [d]\{e_{\mathrm{i}}\}$$

or, writing

$$\{\sigma\}_{\mathrm{t}} = [d]\{e_{\mathrm{t}}\} \tag{8.14}$$

we have

$$\{\sigma\} = [d]\{\epsilon\} - \{\sigma\}_{\mathrm{t}} - [d]\{e_{\mathrm{i}}\} \tag{8.15}$$

Clearly, Eq. (8.15) is the general stress-strain relationship for a three-dimensional isotropic and homogeneous body. From Eqs. (8.5), (8.11) and (8.14) we can show that for a three-dimensional thermally isotropic body, the thermal stress vector is

$$\{\sigma\}_{\mathrm{t}} = \frac{E\alpha T}{(1 - 2\nu)}\{1 \quad 1 \quad 1 \quad 0 \quad 0 \quad 0\}^{\mathrm{T}}$$

A large category of problems in elasticity can be treated as essentially two-dimensional, when the geometry of the body and the forces acting on it are such that identical deformations occur on planes normal to one of the coordinate axes, which can therefore be disregarded. Obviously, this results in a considerable simplification of the mathematical aspect of the solution. These problems are either of the 'plane strain' or the 'plane stress' type.

8.5 Plane strain

Consider the long and uniform slab of Fig. 8.4a, which carries a uniformly distributed 'transverse' load over its entire surface area. Let us now consider the section $A - A$, shown in Fig. 8.4b, which is sufficiently far from the ends $z = 0$ and $z = L$. Then, the slab being loaded uniformly, displacements along z at $A - A$ will be negligible. Furthermore, as $A - A$ is sufficiently remote from the ends, the deformations within it can be assumed to be independent of the z axis, and therefore identical to those of the other sections in the vicinity of and parallel to $A - A$. We therefore conclude that for sections such as $A - A$ it is reasonable to assume

$$\frac{\partial}{\partial z} = w = 0 \tag{8.16a}$$

and as a result,

$$e_{zz} = e_{yz} = e_{zx} = 0 \tag{8.16b}$$

These equations indicate that the state of stress at $A - A$ can be determined by disregarding the z axis. The problem thus reduces to a two-dimensional one in which the displacement and stress patterns are the same at all sections

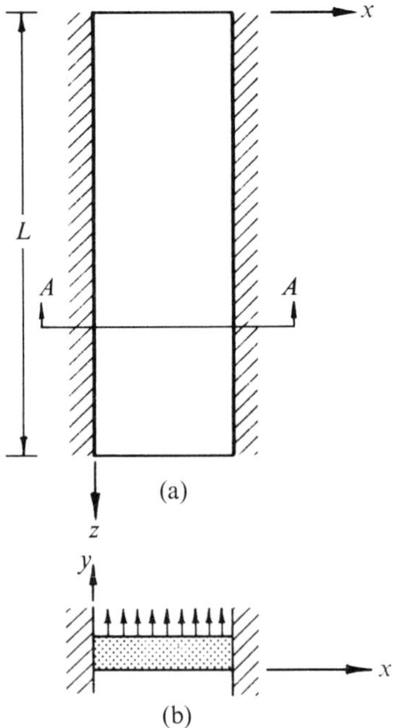

Fig. 8.4 (a) A long slab of uniform thickness, carrying a uniformly distributed transverse load over its entire surface area. (b) Section at A-A.

parallel to $A - A$, except those near the ends where the stress distribution is essentially three-dimensional.

A state of 'plane strain' is said to exist in a body whenever Eqs. (8.16) are satisfied. Long gravity dams, strip footings, long roller bearings etc. are familiar examples of plane strain.

Eq. (8.8c) shows that a consequence of $e_{zz} = 0$ is

$$\sigma_{zz} = \nu(\sigma_{xx} + \sigma_{yy}) \tag{8.17}$$

Clearly, σ_{zz} here is the normal stress required to resist elongation along the z axis.

The constitutive equation for plane strain can be obtained from Eq. (8.15) merely by deleting the rows and columns which refer to the zero strains of Eq. (8.16b). Thus, with $\{\epsilon\}$ as given by Eq. (8.4a),

$$\begin{Bmatrix} \sigma_{xx} \\ \sigma_{yy} \\ \tau_{xy} \end{Bmatrix} = [d]\{\epsilon\} - \{\sigma\}_t - [d]\begin{Bmatrix} e_{ixx} \\ e_{iyy} \\ e_{ixy} \end{Bmatrix} \tag{8.18}$$

in which $[d] = \begin{bmatrix} \lambda + 2G & \lambda & 0 \\ \lambda & \lambda + 2G & 0 \\ 0 & 0 & G \end{bmatrix}$ and $\{\sigma\}_t = \dfrac{E\alpha T}{(1 - 2\nu)}\begin{Bmatrix} 1 \\ 1 \\ 0 \end{Bmatrix}$

Notice that σ_{zz} is not included in Eq. (8.18) since it is linearly dependent on the other normal stresses in accordance with Eq. (8.17).

Problems in which e_{zz} is a non-zero constant are referred to as those of 'generalised' plane strain.

8.6 Plane stress

The method of 'plane stress' can be used to analyze elasticity problems in which the body is subjected to 'in-plane' forces* and one of its dimensions is very small compared to the rest. A typical problem is shown in Fig. 8.5; as the faces $z = \pm t/2$ are free of applied forces,

$$\sigma_{zz} = \tau_{yz} = \tau_{zx} = 0$$

on these faces. Further, since t is sufficiently small compared with the other dimensions of the body, and provided that the forces applied at the edges are uniform across t, we may assume that the above condition is also valid within the cross section of the body. Without substantial loss of accuracy, we may also assume that the remaining stresses, σ_{xx}, σ_{yy} and τ_{xy}, do not vary across t.

* A 'in-plane' force is one that lies on the plane of a thin body.

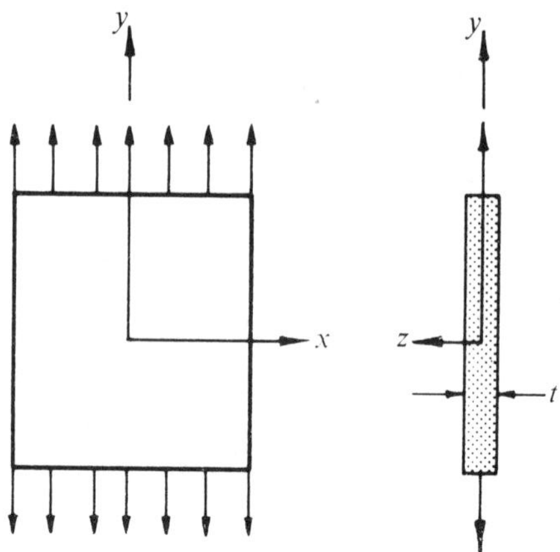

Fig. 8.5 A typical plane stress problem.

The problem thus becomes a two-dimensional one in which the z axis can be disregarded.

On setting $\sigma_{zz} = 0$ into Eqs. (8.8),

$$e_{xx} = \frac{1}{E}(\sigma_{xx} - \nu\sigma_{yy})$$

and

$$e_{yy} = \frac{1}{E}(\sigma_{yy} - \nu\sigma_{xx})$$

Also, the only non-zero shear strain here is (Eq. 8.9a)

$$e_{xy} = \frac{1}{G}\tau_{xy}$$

We shall now write these three equations in the matrix form as

$$\begin{Bmatrix} e_{xx} \\ e_{yy} \\ e_{xy} \end{Bmatrix} = \frac{1}{E}\begin{bmatrix} 1 & -\nu & 0 \\ -\nu & 1 & 0 \\ 0 & 0 & 2(1+\nu) \end{bmatrix}\begin{Bmatrix} \sigma_{xx} \\ \sigma_{yy} \\ \tau_{xy} \end{Bmatrix}$$

By inverting this equation and introducing into the result the thermal and initial strains in accordance with Eq. (8.15), we have

$$\begin{Bmatrix} \sigma_{xx} \\ \sigma_{yy} \\ \tau_{xy} \end{Bmatrix} = [d]\{\epsilon\} - \{\sigma\}_t - [d]\begin{Bmatrix} e_{ixx} \\ e_{iyy} \\ e_{ixy} \end{Bmatrix} \tag{8.19}$$

in which $\{\epsilon\}$ is defined by Eq. (8.4a),

$$[d] = \frac{E}{(1 - \nu^2)} \begin{bmatrix} 1 & \nu & 0 \\ \nu & 1 & 0 \\ 0 & 0 & \dfrac{1 - \nu}{2} \end{bmatrix} \quad \text{and} \quad \{\sigma\}_t = \frac{E\alpha T}{(1 - \nu)} \begin{Bmatrix} 1 \\ 1 \\ 0 \end{Bmatrix}$$

Eq. (8.19) is therefore the constitutive equation for plane stress problems. Since $\sigma_{zz} = 0$, Eq. (8.8c) shows that the strain along z is

$$e_{zz} = -\frac{\nu}{E}(\sigma_{xx} + \sigma_{yy}) \tag{8.20}$$

If the forces applied at the edges are not uniform throughout the thickness of the plate as required in plane stress analysis, but symmetrical about the middle plane at $z = 0$, then we could still use the above procedure by replacing the stresses and strains by their values averaged across the thickness of the plate. Such a problem is said to be one of 'generalized' plane stress.

8.7 The equilibrium equations

Consider the two-dimensional element of Fig. 8.6 which has been isolated from within a body and has unit thickness. The internal stresses shown here represent the effect of the removed parts of the body upon this element. In addition to these stresses we must also consider, for the sake of generality, the so called 'body' and 'inertia' forces which may act on the element.

Gravitational and centrifugal forces are typical examples of 'body' forces to which the body may be subject. Defining $\bar{X}$, $\bar{Y}$ and $\bar{Z}$ as the body force 'accelerations' along the x, y and z axes respectively, we have in the case of this element

$$\text{body force along } x = \rho\bar{X}\Delta x\Delta y$$

and

$$\text{body force along } y = \rho\bar{Y}\Delta x\Delta y$$

where ρ denotes the mass density of the material of the body.

The 'inertia' forces, on the other hand, come into play only when the body is subjected to elastic vibrations; the particles of the body are now excited into a vibratory motion, and according to Newton's second law, it is obvious that in the case of Fig. 8.6

$$\text{inertia force along } x = -\rho\frac{\partial^2 u}{\partial t^2}\Delta x\Delta y$$

Similarly for the y axis. The negative sign above indicates the reactive nature of inertia forces while t is time.

For the element of Fig. 8.6 to be in equilibrium, the vector sum of all forces

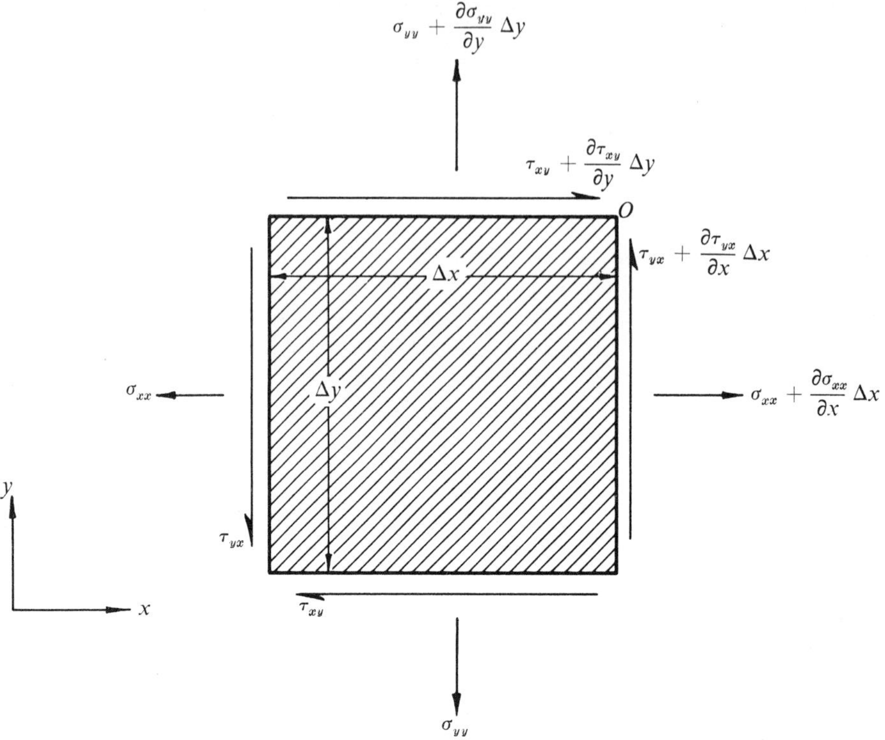

Fig. 8.6 Stresses in a two dimensional differential element.

acting along any arbitrary direction must vanish; so must the moments of all forces about any arbitrary point. Thus, summing along x all the stress resultants, including body and inertia forces, we can show that

$$\left(\frac{\partial \sigma_{xx}}{\partial x} + \frac{\partial \tau_{xy}}{\partial y} + \rho \bar{X} - \rho \frac{\partial^2 u}{\partial t^2}\right)\Delta x \Delta y = 0$$

or, dividing both sides by $\Delta x \Delta y$, we obtain

$$\frac{\partial \sigma_{xx}}{\partial x} + \frac{\partial \tau_{xy}}{\partial y} + \rho \bar{X} = \rho \frac{\partial^2 u}{\partial t^2}$$

We can demonstrate in this way that for a three-dimensional body the above equation becomes

$$\frac{\partial \sigma_{xx}}{\partial x} + \frac{\partial \tau_{xy}}{\partial y} + \frac{\partial \tau_{xz}}{\partial z} + \rho \bar{X} = \rho \frac{\partial^2 u}{\partial t^2} \tag{8.21a}$$

Similarly, for the y and z axes the following equations obtain respectively

$$\frac{\partial \tau_{yx}}{\partial x} + \frac{\partial \sigma_{yy}}{\partial y} + \frac{\partial \tau_{yz}}{\partial z} + \rho \bar{Y} = \rho \frac{\partial^2 v}{\partial t^2} \tag{8.21b}$$

and

$$\frac{\partial \tau_{zx}}{\partial x} + \frac{\partial \tau_{zy}}{\partial y} + \frac{\partial \sigma_{zz}}{\partial z} + \rho \bar{Z} = \rho \frac{\partial^2 w}{\partial t^2} \tag{8.21c}$$

If, for example, the only body force present is that due to gravity, then in accordance with our coordinate system we shall have $\bar{X} = \bar{Z} = 0$ and $\bar{Y} = -g$, where g is acceleration due to gravity. It is also obvious that if the forces acting on the body are static, then the right hand sides of Eqs. (8.21) would vanish.

The moment equilibrium conditions are established similarly by equating to zero the moments of all forces about any point. Assuming that the body forces act at the centre of area of the element and that only static forces are applied to the body, we now equate to zero the sum of all moments about O (Fig. 8.6); this leads to

$$\left(-\tau_{xy} + \tau_{yx} + \rho \bar{X} \frac{\Delta y}{2} - \rho \bar{Y} \frac{\Delta x}{2}\right) \Delta x \Delta y = 0$$

in which anti-clockwise moments have been treated as positive. As Δx and Δy are infinitesimal, the above equation shows that the contribution of the body forces is negligible. It follows therefore that

$$\tau_{xy} = \tau_{yx}$$

We can also demonstrate in this way that

$$\tau_{yz} = \tau_{zy} \quad \text{and} \quad \tau_{zx} = \tau_{xz}$$

As each of these stresses is related linearly (Eqs. 8.9) to its corresponding strain, similar relationships also hold between the shear strains; i.e.,

$$e_{xy} = e_{yx}, \qquad e_{yz} = e_{zy} \quad \text{and} \quad e_{zx} = e_{xz}$$

8.8 Compatibility conditions

Eqs. (8.1) and (8.3) show that if the displacement at a point within the body is known, then all the six components of strain at that point can be obtained explicitly from the derivatives of u, v and w. The fact that six strain components are expressible in terms of three displacement components, suggests that a functional relationship must exist between the various strain components. Indeed, if such relationships did not exist and as a result if it were possible to assign arbitrary values to the strain components, then it is easy to visualize from Fig. 8.2 that continuity of displacements would not exist even though the equilibrium conditions may be satisfied. These inter-relationships between the strain components are in fact the compatibility conditions, which when satisfied,

ensure the continuity of displacements within the deformed body. These conditions are

$$\frac{\partial^2 e_{xx}}{\partial y^2} + \frac{\partial^2 e_{yy}}{\partial x^2} = \frac{\partial^2 e_{xy}}{\partial x\,\partial y} \tag{8.22a}$$

$$\frac{\partial^2 e_{yy}}{\partial z^2} + \frac{\partial^2 e_{zz}}{\partial y^2} = \frac{\partial^2 e_{yz}}{\partial y\,\partial z} \tag{8.22b}$$

$$\frac{\partial^2 e_{zz}}{\partial x^2} + \frac{\partial^2 e_{xx}}{\partial z^2} = \frac{\partial^2 e_{zx}}{\partial z\,\partial x} \tag{8.22c}$$

$$\frac{\partial}{\partial z}\left(\frac{\partial e_{yz}}{\partial x} + \frac{\partial e_{zx}}{\partial y} - \frac{\partial e_{xy}}{\partial z}\right) = 2\frac{\partial^2 e_{zz}}{\partial x\,\partial y} \tag{8.22d}$$

$$\frac{\partial}{\partial x}\left(\frac{\partial e_{zx}}{\partial y} + \frac{\partial e_{xy}}{\partial z} - \frac{\partial e_{yz}}{\partial x}\right) = 2\frac{\partial^2 e_{xx}}{\partial y\,\partial z} \tag{8.22e}$$

and

$$\frac{\partial}{\partial y}\left(\frac{\partial e_{xy}}{\partial z} + \frac{\partial e_{yz}}{\partial x} - \frac{\partial e_{zx}}{\partial y}\right) = 2\frac{\partial^2 e_{yy}}{\partial z\,\partial x} \tag{8.22f}$$

All the above equations can be derived by manipulating Eqs. (8.1) and (8.3). It may be pointed out here that these equations, like the equilibrium equations (Eqs. 8.21), are not directly used in the finite element method.

8.9 Boundary conditions

At the surface of the body the internal stresses must be in equilibrium with the external forces. Let the infinitesimal length AB in Fig. 8.7, which is at the surface

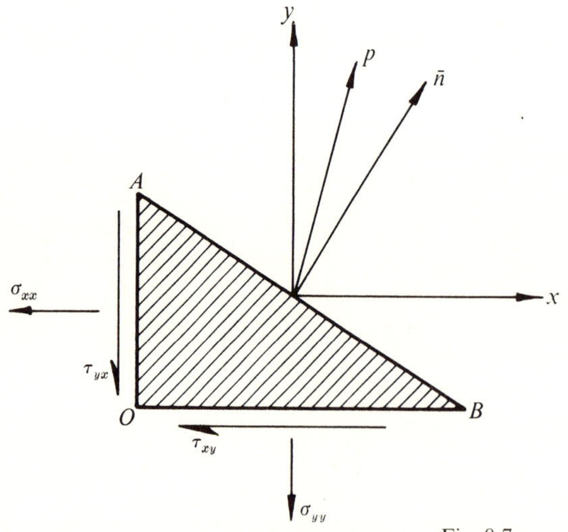

$OA = AB \cos(\bar{n}, x)$
$OB = AB \cos(\bar{n}, y)$
$\bar{n}$ is normal to AB

$(\bar{n}, x)$ = angle between $\bar{n}$ and x
$(\bar{n}, y)$ = angle between $\bar{n}$ and y

Fig. 8.7

of the body, be subjected to the 'total stress' p, and let p_x and p_y be the components of p along the x and y axes respectively. As OAB is in equilibrium, it is clear from this figure that

$$p_x AB = \sigma_{xx} OA + \tau_{xy} OB$$

and

$$p_y AB = \tau_{yx} OA + \sigma_{yy} OB$$

or

$$p_x = \sigma_{xx} \cos(\bar{n}, x) + \tau_{xy} \cos(\bar{n}, y)$$

and

$$p_y = \tau_{yx} \cos(\bar{n}, x) + \sigma_{yy} \cos(\bar{n}, y)$$

in which $(\bar{n}, x)$ and $(\bar{n}, y)$ are the angles between the outward normal $\bar{n}$ and the x and y axes respectively. This analysis is easily extended to demonstrate that in three dimensions, every point on the surface of the body must satisfy the following 'boundary' condition

$$\begin{Bmatrix} p_x \\ p_y \\ p_z \end{Bmatrix} = \begin{bmatrix} \sigma_{xx} & \tau_{xy} & \tau_{xz} \\ \tau_{yx} & \sigma_{yy} & \tau_{yz} \\ \tau_{zx} & \tau_{zy} & \sigma_{zz} \end{bmatrix} \begin{Bmatrix} l \\ m \\ n \end{Bmatrix} \tag{8.23}$$

where l, m and n are the direction cosines of the normal $\bar{n}$ with respect to the coordinate system to which Eq. (8.23) refers. The two-dimensional array in the above equation is called the 'stress tensor', in this case for the point at which p is applied.

In the Theory of Elasticity p is also called the 'stress vector', defined as the resultant force per unit area exerted across ΔS upon the material within V by the material outside V. Here ΔS is an element of the surface S which encloses an arbitrary volume V of the body.

8.10 Principal stresses

Imagine that OAB in Fig. 8.7 is located inside the body, rather than at its surface. Also, let p be the principal stress whose principal plane is AB. As a principal stress is normal to its principal plane, the direction of p and that of the normal $\bar{n}$ now coincide. Consequently,

$$p_x = p \cos(\bar{n}, x)$$

and

$$p_y = p \cos(\bar{n}, y)$$

Similarly, in three dimensions

$$p_x = pl, \qquad p_y = pm \quad \text{and} \quad p_z = pn$$

Substituting these into Eq. (8.23) and reorganising, we have

$$\begin{bmatrix} (\sigma_{xx} - p) & \tau_{xy} & \tau_{xz} \\ \tau_{yx} & (\sigma_{yy} - p) & \tau_{yz} \\ \tau_{zx} & \tau_{zy} & (\sigma_{zz} - p) \end{bmatrix} \begin{Bmatrix} l \\ m \\ n \end{Bmatrix} = \begin{Bmatrix} 0 \\ 0 \\ 0 \end{Bmatrix} \tag{8.24}$$

If the stress tensor refers to a particular point within the body, then the principal stresses and their directions at that point can now be determined by solving Eq. (8.24).

In numerous elasticity problems, it is necessary to compute the principal stresses and their orientations at various points. We shall therefore discuss here in some detail the implications and methods of solution of Eq. (8.24).

By inspection, we may say that a solution of Eq. (8.24) is

$$l = m = n = 0$$

This solution, which is termed as the 'trivial' solution, is however not acceptable, as it violates the condition

$$l^2 + m^2 + n^2 = 1 \tag{8.25}$$

which must hold amongst the direction cosines.* In order to determine the 'non-trivial' solution of Eq. (8.24), we now set its determinant to zero, upon which the following cubic equation in p results:

$$p^3 - p^2\text{I} + p\text{II} - \text{III} = 0 \tag{8.26}$$

in which

$$\text{I} = (\sigma_{xx} + \sigma_{yy} + \sigma_{zz})$$

$$\text{II} = (\sigma_{xx}\sigma_{yy} + \sigma_{yy}\sigma_{zz} + \sigma_{zz}\sigma_{xx} - \tau_{xy}^2 - \tau_{yz}^2 - \tau_{zx}^2)$$

and

$$\text{III} = (\sigma_{xx}\sigma_{yy}\sigma_{zz} + 2\tau_{xy}\tau_{yz}\tau_{zx} - \sigma_{xx}\tau_{yz}^2 - \sigma_{yy}\tau_{xz}^2 - \sigma_{zz}\tau_{xy}^2)$$

I, II and III are called the first, second and third 'invariants' respectively of the stress tensor. An 'invariant' quantity is one whose value is not affected by a coordinate transformation.

The three roots, p_1, p_2 and p_3 of Eq. (8.26) are the required principal stress values. The orientation of each of these stresses can be determined as follows:

Let the direction cosines of p_1, for example, be l_1, m_1 and n_1. Substituting p_1 and these into Eq. (8.24), we have

$$\begin{aligned}
(\sigma_{xx} - p_1)l_1 + \tau_{xy}m_1 + \tau_{xz}n_1 &= 0 \\
\tau_{yx}l_1 + (\sigma_{yy} - p_1)m_1 + \tau_{yz}n_1 &= 0 \\
\tau_{zx}l_1 + \tau_{zy}m_1 + (\sigma_{zz} - p_1)n_1 &= 0
\end{aligned}$$

The above equations (together with Eq. (8.25) if necessary) can now be solved for l_1, m_1 and n_1. The directions of p_2 and p_3 can also be found in this way.

In practice, however, it is often more convenient to determine the principal stresses and their directions as the roots and vectors respectively of the following

* We have assumed here that the direction cosines are always real. This is true, as the stress tensor is real and symmetrical.

'eigenvalue' equation into which Eq. (8.24) can be rearranged:

$$\begin{bmatrix} \sigma_{xx} & \tau_{xy} & \tau_{xz} \\ \tau_{yx} & \sigma_{yy} & \tau_{yz} \\ \tau_{zx} & \tau_{zy} & \sigma_{zz} \end{bmatrix} \begin{Bmatrix} l \\ m \\ n \end{Bmatrix} = p \begin{Bmatrix} l \\ m \\ n \end{Bmatrix} \tag{8.27}$$

The roots (eigenvalues) and vectors (eigenvectors) of Eq. (8.27) are easily computed by using standard 'subroutines' which are built into most large computers. It is important to point out, however, that unless 'normalized', the vectors of Eq. (8.27) will be proportional but not equal to the true direction cosines.* For instance, let the vector of Eq. (8.27) corresponding to the root p_1 be $\{l'_1 \; m'_1 \; n'_1\}$. Now, as the true direction cosines are proportional to this vector, we may write

$$\begin{Bmatrix} l_1 \\ m_1 \\ n_1 \end{Bmatrix} = r_1 \begin{Bmatrix} l'_1 \\ m'_1 \\ n'_1 \end{Bmatrix}$$

Here, r_1, defined as

$$r_1 = \frac{1}{(l_1'^2 + m_1'^2 + n_1'^2)^{\frac{1}{2}}}$$

may be called the 'normalizing' factor for p_1. The factors, r_2 and r_3 can be computed similarly to give the true directions of p_2 and p_3 respectively. In plane problems Eq. (8.27) will obviously reduce to

$$\begin{bmatrix} \sigma_{xx} & \tau_{xy} \\ \tau_{yx} & \sigma_{yy} \end{bmatrix} \begin{Bmatrix} l \\ m \end{Bmatrix} = p \begin{Bmatrix} l \\ m \end{Bmatrix}$$

in which we have assumed the body to lie on the x–y plane.

8.11 The Lamé equations

In classical elasticity there are two different ways in which we can solve a given problem.

(1) We can take the six components of stress (Eq. 8.13) as the basic unknowns, and determine them by solving the stress equilibrium (Eqs. 8.21) and compatibility (Eqs. 8.22) equations.

(2) Alternatively, the three components of displacement can be treated as the basic unknowns, to be determined by solving the stress equilibrium equations.

* 'Normalization' in this context simply means adjustment of the values of l'_1, m'_1 and n'_1, so that they satisfy the condition of Eq. (8.25) after adjustment.

In both cases, the solution must also satisfy the prescribed boundary conditions. The second method is usually preferred, as the number of unknowns here is half that of the first. Indeed, as the reader must have realised by now, the finite element method is fundamentally one of 'unknown displacements'.

In order to solve for the displacements, we must now express the stress equilibrium equations in terms of displacements. This can be done as follows: By multiplying out the first row of Eq. (8.11), we have

$$\sigma_{xx} = (\lambda + 2G)e_{xx} + \lambda e_{yy} + \lambda e_{zz}$$

or, with

$$\bar{e} = (e_{xx} + e_{yy} + e_{zz}), \tag{8.28}$$

$$\sigma_{xx} = \lambda \bar{e} + 2Ge_{xx} = \lambda \bar{e} + 2G\frac{\partial u}{\partial x}$$

Also from Eqs. (8.3) and (8.9),

$$\tau_{xy} = G\left(\frac{\partial u}{\partial y} + \frac{\partial v}{\partial x}\right) \quad \text{and} \quad \tau_{xz} = G\left(\frac{\partial u}{\partial z} + \frac{\partial w}{\partial x}\right)$$

By substituting these expressions for σ_{xx}, τ_{xy} and τ_{xz} into Eq. (8.21a), we can show that

$$\lambda\frac{\partial \bar{e}}{\partial x} + G\left(\frac{\partial^2 u}{\partial x^2} + \frac{\partial^2 u}{\partial y^2} + \frac{\partial^2 u}{\partial z^2}\right) + G\left(\frac{\partial^2 u}{\partial x^2} + \frac{\partial^2 v}{\partial x\,\partial y} + \frac{\partial^2 w}{\partial x\,\partial z}\right) + \rho\bar{X} = \rho\frac{\partial^2 u}{\partial t^2}$$

or,

$$\lambda\frac{\partial \bar{e}}{\partial x} + G\left(\frac{\partial^2 u}{\partial x^2} + \frac{\partial^2 u}{\partial y^2} + \frac{\partial^2 u}{\partial z^2}\right) + G\frac{\partial}{\partial x}\left(\frac{\partial u}{\partial x} + \frac{\partial v}{\partial y} + \frac{\partial w}{\partial z}\right) + \rho\bar{X} = \rho\frac{\partial^2 u}{\partial t^2}$$

or, from Eq. (8.28) and with the Laplacian operator ∇^2, defined as

$$\nabla^2 = \frac{\partial^2}{\partial x^2} + \frac{\partial^2}{\partial y^2} + \frac{\partial^2}{\partial z^2}$$

we have

$$(\lambda + G)\frac{\partial \bar{e}}{\partial x} + G\nabla^2 u + \rho\bar{X} = \rho\frac{\partial^2 u}{\partial t^2} \tag{8.29a}$$

Similarly, we can show that

$$(\lambda + G)\frac{\partial \bar{e}}{\partial y} + G\nabla^2 v + \rho\bar{Y} = \rho\frac{\partial^2 v}{\partial t^2} \tag{8.29b}$$

and

$$(\lambda + G)\frac{\partial \bar{e}}{\partial z} + G\nabla^2 w + \rho\bar{Z} = \rho\frac{\partial^2 w}{\partial t^2} \tag{8.29c}$$

These equations, which are due to Lamé (hence their name) represent a synthesis of the equilibrium conditions and the geometrical and material

aspects of the body. They can now be solved with prescribed boundary conditions for displacements.

When the geometrical configuration of the body is complex, as it usually is in most two and three dimensional practical engineering problems, the closed form solution of Eqs. (8.29) becomes extremely difficult and nearly always impossible. The only alternative, under these circumstances, is to employ numerical methods of solution.

9 General element characteristics

9.1 The unit displacement theorem

This theorem[1] is perhaps the simplest tool available for deriving the stiffness characteristics of finite elements.* It establishes, *via* the principle of virtual work, a functional relationship between the forces required to maintain the equilibrium of a body and the true stress distribution within it. The theorem can be derived as follows:

Let the single force F_k act at the surface point k of a three-dimensional elastic body (Fig. 9.1) for which assumptions 1 and 2 of section 2.1 are valid. The body is in equilibrium and let $\{\sigma\}$ and $\{\epsilon\}$ denote the stresses and strains within it. Also, let u_k be the displacement of k along the direction of F_k.

Now, if we give k a small virtual displacement of δu_k along F_k, then this will obviously alter all the strain components everywhere within the body. Typically, let the change in e_{xx} be δe_{xx}. Since any alteration of strain alters the total strain energy content of the body, it is clear from Fig. 9.2 that the change in strain energy 'per unit volume'† due to δe_{xx} will be equal to $\sigma_{xx}\,\delta e_{xx}$. Repeating this process for all the strain components in turn, it is obvious that the total change of strain energy of the body due to δu_k is

$$\Delta H = \int_V (\sigma_{xx}\,\delta e_{xx} + \sigma_{yy}\,\delta e_{yy} + \sigma_{zz}\,\delta e_{zz} + \tau_{xy}\,\delta e_{xy} + \tau_{yz}\,\delta e_{yz} + \tau_{zx}\,\delta e_{zx})\,\mathrm{d}v \tag{9.1}$$

where the integration is to be performed over the entire volume of the body. Writing the integrand of Eq. (9.1) compactly as the 'scalar product' $\{\delta\epsilon\}^{\mathrm{T}} . \{\sigma\}$, we can express this equation also as

$$\Delta H = \int_V \{\delta\epsilon\}^{\mathrm{T}} . \{\sigma\}\,\mathrm{d}v \tag{9.2}$$

in which $\{\delta\epsilon\}^{\mathrm{T}} = \{\delta e_{xx}\ \delta e_{yy} \cdots \delta e_{zx}\}$

and $\{\sigma\} = \{\sigma_{xx}\ \sigma_{yy}\ \sigma_{zz} \cdots \tau_{zx}\}^{\mathrm{T}}$

As the body deforms in a geometrically linear fashion (section 2.1), a linear

* See section 14.2 for an alternative approach.

† It is proved in elementary strength of materials that the area under the stress-strain curve represents strain energy per unit volume of the body.

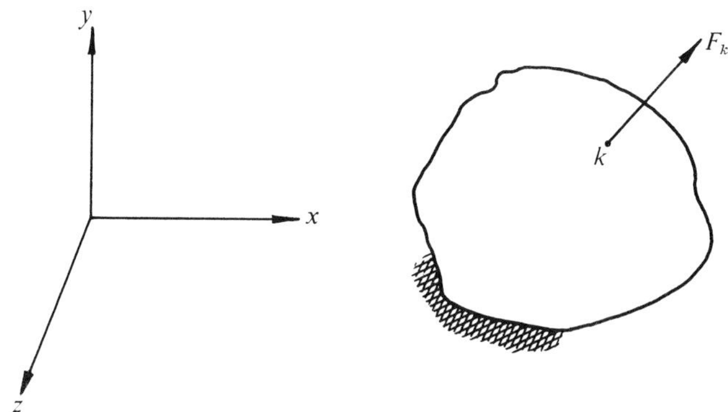

Fig. 9.1 A three-dimensional elastic body subjected to a force F_k acting at the surface point k.

relationship will prevail between the strains and the displacements. In particular, let

$$\{\epsilon\} = \{f\}u_k \tag{9.3}$$

where $\{f\}$ represents the linear relationship between the strains and u_k and $\{\epsilon\}$ is the strain vector defined by Eq. (8.4). Differentiating both sides of Eq. (9.3), we have

$$\{\delta\epsilon\} = \{f\}\,\delta u_k \tag{9.4}$$

Then, eliminating $\{\delta\epsilon\}$ between Eqs. (9.2) and (9.4),

$$\Delta H = \int_V \delta u_k \,.\, \{f\}^{\mathrm{T}} \,.\, \{\sigma\}\, \mathrm{d}v$$

Now, assuming that no energy is dissipated in any way during deformation, it is clear that ΔH must be equal to the virtual work done by F_k in moving

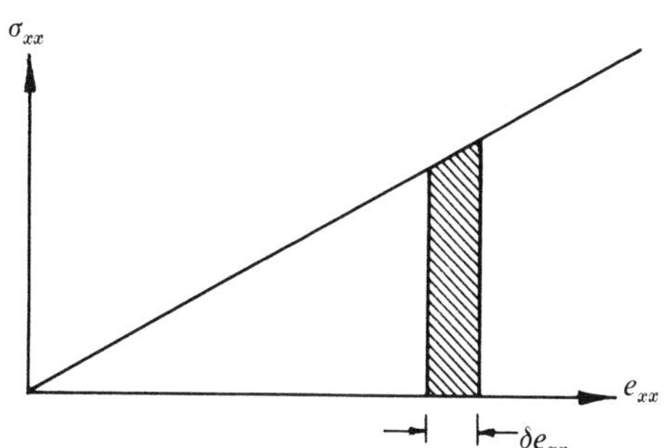

Fig. 9.2 The $\sigma_{xx} - e_{xx}$ diagram for the body of Fig. 9.1.

through δu_k; that is

$$\delta u_k \,.\, F_k = \int_V \delta u_k \,.\, \{f\}^{\mathrm{T}} \,.\, \{\sigma\}\, \mathrm{d}v$$

Since δu_k is arbitrary, it is now cancelled out to give

$$F_k = \int_V \{f\}^{\mathrm{T}} \,.\, \{\sigma\}\, \mathrm{d}v \tag{9.5}$$

Equation (9.5) states the 'unit displacement Theorem' in its simplest form in which a single force is considered. Its general form for multiple forces is easily obtained as follows:

Instead of a single force, let the forces $F_1, F_2 \cdots F_N$ act at the surface points $1, 2 \cdots N$ of the body and let $u_1, u_2 \cdots u_N$ be the respective displacements of these points along $F_1, F_2 \cdots F_N$. Then, it is clear that an equation similar to Eq. (9.5) will obtain for each force. Writing

$$\{P\} = \{F_1 \;\; F_2 \cdots F_N\}^{\mathrm{T}}$$

we can subsequently show that in the case of multiple forces the counterpart of Eq. (9.5) will be

$$\{P\} = \int_V [f]^{\mathrm{T}} \,.\, \{\sigma\}\, \mathrm{d}v \tag{9.6}$$

in which matrix $[f]$ is defined by

$$\{\epsilon\} = [f]\{\bar{\delta}\} \tag{9.7}$$

where $$\{\bar{\delta}\} = \{u_1 \;\; u_2 \cdots u_N\}^{\mathrm{T}}$$

Equation (9.6) is the mathematical statement of the unit displacement Theorem.

9.2 Element stiffness characteristics

Equation (9.6) provides the basis from which the stiffness characteristics of any finite element can be derived. When applying this equation to a given element, $\{P\}$ will be treated as a vector listing all the nodal forces of that element. Furthermore, when the element is considered in isolation, $\{P\}$ must be equal to the extremity (internal) forces of the element. Denoting the extremity forces by the vector $\{\bar{P}\}$, we therefore have from Eq. (9.6)

$$\{\bar{P}\} = \int_V [f]^{\mathrm{T}} \,.\, \{\sigma\}\, \mathrm{d}v \tag{9.8}$$

It is not necessary at this stage to specify the element's dimension, number of extremities or degrees of freedom; we merely state that the vector $\{\bar{P}\}$ lists all

its extremity forces while $\{\bar{\delta}\}$ lists all its extremity displacements. For example, if the element in question is that of a rigid jointed space frame, then $\{\bar{P}\}$ and $\{\bar{\delta}\}$ will respectively denote the left and right hand vectors of Eq. (6.15).

Now, eliminating $\{\epsilon\}$ between Eqs. (9.7) and (8.15), we have

$$\{\sigma\} = [d][f]\{\bar{\delta}\} - \{\sigma\}_t - [d]\{e_i\} \tag{9.9}$$

Finally, elimination of $\{\sigma\}$ between Eqs. (9.8) and (9.9) leads to

$$\{\bar{P}\} = \int_V [f]^T \cdot [d] \cdot [f] \, dv \{\bar{\delta}\} - \int_V [f]^T \cdot \{\sigma\}_t \, dv - \int_V [f]^T \cdot [d] \cdot \{e_i\} \, dv \tag{9.10}$$

$\{\bar{\delta}\}$ has been placed outside the integrand since it is not a function of the current coordinates. In the absence of thermal and initial stresses the first integral of Eq. (9.10) obviously relates the extremity forces of the element to its extremity displacements. Therefore, by the definition of section 2.5, we have

$$[K_i] = \int_V [f]^T \cdot [d] \cdot [f] \, dv \tag{9.11}$$

If $[f]$ for the element is known, then the element stiffness matrix $[K_i]$ can be derived by evaluating the integral of Eq. (9.11).

The second and third integrals of Eq. (9.10) respectively represent the thermal and initial strain forces. Denoting these by

$$\{P_t\} = \int_V [f]^T \cdot \{\sigma\}_t \, dv \qquad \text{(thermal)} \tag{9.12}$$

and

$$\{P_i\} = \int_V [f]^T \cdot [d] \cdot \{e_i\} \, dv \quad \text{(initial strain)} \tag{9.13}$$

we can write Eq. (9.10) also as

$$[K_i]\{\bar{\delta}\} = \{\bar{P}\} + \{P_t\} + \{P_i\} \tag{9.14}$$

As explained in Chapter 3, the $[K_i]$'s of individual elements representing a given body will be assembled to form the overall stiffness matrix for the entire body. However, if we carry out this assembly starting with Eq. (9.14), then the $\{P_t\}$'s and $\{P_i\}$'s would appear on the right hand side of Eq. (3.9). This clearly indicates that thermal and initial strain forces, when present, must be treated as external forces acting at the nodes to which the element is attached, in addition to all other external forces which may be active at these nodes.

9.3 Distributed external forces

We saw in section 5.2 that the distributed external forces on a beam element could be replaced by the vector $\{P_e\}$, which represents equivalent concentrated

external forces, to be applied at the nodes to which the element is attached. The general formula for calculating $\{P_e\}$ for any finite element can be derived as follows:

Let function $\{U\}$ define the displacements everywhere within the element, and let it be related to the element's extremity displacements as

$$\{U\} = [b]\{\bar{\delta}\} \tag{9.15}$$

in which matrix $[b]$ is a function of the current coordinates only. Also, let p, which may be a constant or a function of the current coordinates, denote the intensity of the external distributed surface load acting on the element. Then, the work done by the small concentrated force, $p\ \Delta s$, during a virtual displacement of $\{\delta U\}$ is obviously equal to $\{\delta U\}^{\mathrm{T}} \,.\, p\ \Delta s$; hence, the total virtual work done by the distributed load is

$$\int_S \{\delta U\}^{\mathrm{T}} \,.\, p \,\mathrm{d}s$$

in which the integration is performed over the area of the element on which p is active. This work must be equal to that done by $\{P_e\}$ in moving through the corresponding virtual extremity displacement, $\{\delta\bar{\delta}\}$; that is

$$\{\delta\bar{\delta}\}^{\mathrm{T}} \,.\, \{P_e\} = \int_S \{\delta U\}^{\mathrm{T}} \,.\, p \,\mathrm{d}s \tag{9.16}$$

Differentiating both sides of Eq. (9.15), we have

$$\{\delta U\} = [b]\{\delta\bar{\delta}\} \tag{9.17}$$

Then, following the rule of transpose multiplication ($[AB]^{\mathrm{T}} = [B]^{\mathrm{T}} \,.\, [A]^{\mathrm{T}}$), we can show from Eqs. (9.16) and (9.17) that

$$\{\delta\bar{\delta}\}^{\mathrm{T}} \,.\, \{P_e\} = \int_S \{\delta\bar{\delta}\}^{\mathrm{T}} \,.\, [b]^{\mathrm{T}} \,.\, p \,\mathrm{d}s$$

or, as $\{\delta\bar{\delta}\}$ is arbitrary, it is cancelled out to give

$$\{P_e\} = \int_S [b]^{\mathrm{T}} \,.\, p \,\mathrm{d}s \tag{9.18}$$

The equivalent force vector for any finite element can be calculated from Eq. (9.18). A typical application of this equation will be given in section 9.7.

9.4 Body forces

By definition, the body forces (section 8.7) are associated with the volume of the finite element. Like the distributed forces considered in section 9.3, it is also possible to express body forces in terms of an equivalent external nodal force vector, which we shall denote by $\{P_b\}$.

The derivation of $\{P_b\}$ is identical to that of $\{P_e\}$. In this case we can show that the total virtual work done by the body forces is equal to

$$\int_V \rho\{\delta U\}^{\mathrm{T}} \,.\, \begin{Bmatrix} \bar{X} \\ \bar{Y} \\ \bar{Z} \end{Bmatrix} \mathrm{d}v$$

in which the integration is performed throughout the entire volume of the element. This work is equal to that done by $\{P_b\}$ in moving through $\{\delta\bar{\delta}\}$. Consequently, it is easy to show, following the procedure of the last section, that

$$\{P_b\} = \int_V \rho[b]^{\mathrm{T}} \,.\, \begin{Bmatrix} \bar{X} \\ \bar{Y} \\ \bar{Z} \end{Bmatrix} \mathrm{d}v \tag{9.19}$$

A typical application of this equation will be given in section 11.1.

9.5 The displacement function—accuracy of solution

The fundamental assumption in the finite element method of analysis is that the response of a continuous body to a given set of applied forces is equivalent to that of a system of discrete elements into which the body may be imagined to be subdivided. From the energy point of view, the equivalence between the body and its finite element model would therefore be exact, if the strain energy of the deformed body were equal to that of its discrete model. Eqs. (9.1) and (9.7) indicate that this requirement is fulfilled when the strain distribution within the model is exactly the same as that prevailing within the body.

Exact determination of the strain distribution function, $[f]$, however, presents considerable difficulties, since this can be done only by a rigorous solution of the equations of linear elasticity. In most cases, particularly in two- and three-dimensional elements, solution of these equations in the closed form is very difficult and usually impossible. This compels us to adopt an alternative procedure in which we assume a displacement function, $\{U\}$, for the element. The strain distribution, $[f]$, can then be derived from it by differentiation. The strain energy content of the finite element model and consequently the degree of accuracy and the manner in which the solution converges to its exact value with increasingly finer subdivision of the body, is determined primarily by how closely the assumed displacements (and hence strains) resemble their exact counterparts.

When assuming the displacement function, $\{U\}$, for a finite element which has n extremities, we are guided by the following considerations:

(a) If one or more mutually independent components of displacement (or function) occur within the element, then each component will be assumed

as a polynomial defined by n unknown constants (e.g., Chapters 11 and 12 and section 13.2). If, on the other hand, m displacements are derived from an independent displacement, then the independent displacement will be assumed as a polynomial defined by a total of $n(m + 1)$ unknown constants (e.g., section 9.7b and Chapter 10). Because, in both cases, the values of the unknown constants can then be determined directly in terms of the extremity displacements of the element.

(b) $\{U\}$ must be such that the slopes and displacements within the element are continuous, since any discontinuity would lead to infinite strains and hence infinite strain energy. Furthermore, it is desirable, though not necessary, that the slopes and displacements be also continuous across common interfaces between adjacent two- and three-dimensional elements (inter-element continuity in one-dimensional elements is ensured by the assembly procedure; see section 3.1). This requirement is fulfilled when the displacements and slopes along a common interface depend only on those of the nodes defining that interface.

(c) As we saw in section 8.11, the solution of an elastic problem with displacements as the basic unknowns entails solving the stress-equilibrium equations together with the prescribed boundary conditions. It is desirable therefore that the stresses due to the assumed displacements also satisfy the stress-equilibrium equations (Eqs. 8.21). This additional requirement makes the choice of $\{U\}$ very difficult indeed. A method of complying with this is to assume $\{U\}$ such that the strains, and hence the stresses, remain constant throughout the element. Eqs. (8.21) will then be satisfied identically.

(d) The choice of $\{U\}$ from the point of view of convergence is generally a difficult one and further investigation is needed in this area. An apparently satisfactory procedure[2] is to choose $\{U\}$ such that strains remain constant throughout the element and no straining of the element is caused by its rigid body displacement.

It is important to observe that the assumed displacements of the element will normally refer to its 'local' coordinates (section 6.2). Also, the local axes of all elements representing the body will normally be made parallel to each other. If, as in Chapters 6 and 7, it is not possible to do so, then the 'local' forces, displacements etc. of each element will have to be transformed so that they all refer to a common 'global' system of coordinates. The need for such a transformation obviously arises in bent and/or curved structures. It also arises in anisotropic or orthotropic problems in which the directions of anisotropy or orthotropy do not coincide with the local axes (see section 11.6).

9.6 Subdivision of the body

The criterion of constant strain discussed in section 9.5(d) provides a convenient basis for determining the method of subdivision of the body into a system of finite elements. For instance, if in a certain part of the body the stresses vary

rapidly, then we must provide there a relatively large number of relatively small elements, so that the true stress variation is well approximated. This approximation will clearly tend to the true stress distribution as the subdivision becomes increasingly fine. By the same token, the subdivision may be 'eased off' into progressively larger elements as we move away from this region and proceed towards those where the stresses vary less rapidly. The 'easing off' should be gradual, since experience shows that a sudden or violent change of element size is likely to lead to gross inaccuracies. A scale factor of more than 2 is not recommended except under special circumstances.

If the variation of true stresses or strains within the deformed body cannot be anticipated, then an initial subdivision, based on guesswork, may be used. The computed variation of the stresses or strains due to this subdivision will form the basis for a later improved scheme of subdivision. In practice it is advisable to try out two or more different schemes of subdivision in order to determine how the solution converges to its exact value.

9.7 Some examples

(a) *The bar element*

To begin with, let us calculate the $[K_i]$ matrix of the uniform bar element of Fig. 9.3 by using Eq. (9.11). In bar elements only axial forces and displacements

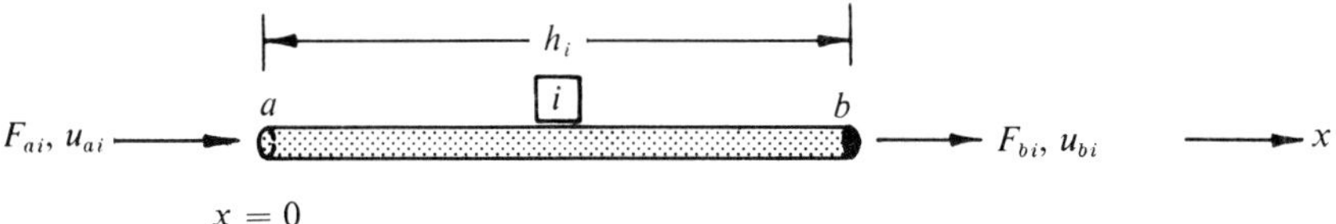

Fig. 9.3 A uniform bar element.

occur. Now, as the element has two extremities, $n = 2$; according to section 9.5(a), the axial displacement $u(x)$ can therefore be assumed as a polynomial defined by two constants; that is

$$U = u(x) = A_0 + A_1 x \tag{9.20}$$

in which A_0 and A_1 are the unknown constants. The values of these constants can be found as follows: From Eq. (9.20) and Fig. 9.3, we have

$$u_{ai} = u(0) = A_0$$

and

$$u_{bi} = u(h_i) = A_0 + A_1 h_i$$

or, in the matrix form,

$$\begin{Bmatrix} u_{ai} \\ u_{bi} \end{Bmatrix} = \begin{bmatrix} 1 & 0 \\ 1 & h_i \end{bmatrix} \begin{Bmatrix} A_0 \\ A_1 \end{Bmatrix} \tag{9.21}$$

Now, noting that by definition

$$\{\bar{\delta}\} = \begin{Bmatrix} u_{ai} \\ u_{bi} \end{Bmatrix}$$

and introducing

$$[c] = \begin{bmatrix} 1 & 0 \\ 1 & h_i \end{bmatrix} \tag{9.22}$$

and

$$\{A\} = \begin{Bmatrix} A_0 \\ A_1 \end{Bmatrix}$$

we can write Eq. (9.21) compactly as

$$\{\bar{\delta}\} = [c]\{A\} \tag{9.23}$$

Consequently, the values of the constants will be given by

$$\{A\} = [c^{-1}]\{\bar{\delta}\} \tag{9.24}$$

[The following observations, which apply to all finite elements, can now be made:

(a) The relationship between the extremity displacements and the constants defining $\{U\}$ will always be expressed in terms of Eq. (9.23).

(b) If the number of unknown constants defining $\{U\}$ is not equal to that prescribed in section 9.5(a), then the $[c]$ matrix will become rectangular making Eq. (9.24) meaningless, and

(c) The $[c]$ matrix depends on the extremity coordinates only and is independent of the current coordinates.]

In case of the bar element, $[d]$ is the single quantity, E_i (= Young's modulus), and

$$\text{strain} = \epsilon = e_{xx} = \frac{du}{dx}$$

Then, differentiating Eq. (9.20), we have

$$\epsilon = A_1 = \{0 \quad 1\}\begin{Bmatrix} A_0 \\ A_1 \end{Bmatrix} = \{0 \quad 1\}\{A\} \tag{9.25}$$

Elimination of $\{A\}$ between Eqs. (9.24) and (9.25) leads to

$$\epsilon = \{0 \quad 1\}[c^{-1}]\{\bar{\delta}\} \tag{9.26}$$

Comparison between Eqs. (9.7) and (9.26) shows that in this case

$$\{f\} = \{0 \quad 1\}[c^{-1}] \tag{9.27}$$

Inverting $[c]$ of Eq. (9.22) we get

$$[c^{-1}] = \begin{bmatrix} 1 & 0 \\ -1/h_i & 1/h_i \end{bmatrix} \tag{9.28}$$

Consequently, Eq. (9.27) gives

$$\{f\} = \{0 \quad 1\}\begin{bmatrix} 1 & 0 \\ -1/h_i & 1/h_i \end{bmatrix} = \{-1/h_i \quad 1/h_i\} \tag{9.29}$$

Therefore, from Eq. (9.11), we have

$$[K_i] = \int_V \begin{Bmatrix} -1/h_i \\ 1/h_i \end{Bmatrix} E_i\{-1/h_i \quad 1/h_i\}\, \mathrm{d}v = (a_i E_i/h_i^2)\int_0^{h_i} \begin{bmatrix} 1 & -1 \\ -1 & 1 \end{bmatrix} \mathrm{d}x \tag{9.30}$$

in which a_i is the area of cross section of the bar. It is a simple matter now to show that

$$[K_i] = k_i\begin{bmatrix} 1 & -1 \\ -1 & 1 \end{bmatrix} \tag{9.31}$$

which agrees exactly with that derived in section 2.6 (also see Eq. (7.2)).

(b) *The beam element*

As a second example let us calculate the $[K_i]$ matrix of the beam element of Fig. 2.6 (axial and torsional modes will be ignored). Since the rotational displacement of the element is derived from its transverse deflection, $m = 1$ according to section 9.5(a). Also, since the element has 2 extremities, $n = 2$. Consequently, the transverse deflection $v(x)$ of the element will be assumed as a polynomial defined by a total of $2(1 + 1) = 4$ unknown constants (section 9.5a); thus, let

$$U = v(x) = A_0 + A_1 x + A_2 x^2 + A_3 x^3 \tag{9.32}$$

represent the element's transverse deflection function in which $A_0 \cdots A_3$ are the unknown constants. By definition, the rotation of the element is therefore

$$\theta(x) = \frac{\mathrm{d}v}{\mathrm{d}x} = A_1 + 2A_2 x + 3A_3 x^2 \tag{9.33}$$

The extremity displacements, v_{ai} and v_{bi}, are now obtained by substituting the coordinates of the extremities into Eq. (9.32). Thus,

$$v_{ai} = v(0) = A_0$$

and

$$v_{bi} = v(h_i) = A_0 + A_1 h_i + A_2 h_i^2 + A_3 h_i^3$$

Similarly, from Eq. (9.33) we have

$$\theta_{ai} = \theta(0) = A_1$$

and

$$\theta_{bi} = \theta(h_i) = A_1 + 2A_2h_i + 3A_3h_i^2$$

The above four equations will now be expressed in the matrix form of Eq. (9.23), by noting that now

$$\{\delta\} = \begin{Bmatrix} u_{ai} \\ \theta_{ai} \\ u_{bi} \\ \theta_{bi} \end{Bmatrix} \quad \text{and} \quad \{A\} = \begin{Bmatrix} A_0 \\ A_1 \\ A_2 \\ A_3 \end{Bmatrix}$$

Then, it may easily be verified that in this case

$$[c] = \begin{bmatrix} 1 & 0 & 0 & 0 \\ 0 & 1 & 0 & 0 \\ 1 & h_i & h_i^2 & h_i^3 \\ 0 & 1 & 2h_i & 3h_i^2 \end{bmatrix} \tag{9.34}$$

whose inverse is

$$[c^{-1}] = \frac{1}{h_i^3}\begin{bmatrix} h_i^3 & 0 & 0 & 0 \\ 0 & h_i^3 & 0 & 0 \\ -3h_i & -2h_i^2 & 3h_i & -h_i^2 \\ 2 & h_i & -2 & h_i \end{bmatrix} \tag{9.35}$$

Since the 'strain' within a beam in bending is defined as its curvature, the $[d]$ matrix of the uniform beam element i becomes simply E_iI_i, and Eq. (9.11) reduces to

$$[K_i] = E_iI_i\int_0^{h_i} \{f\}^{\mathrm{T}}\{f\}\,\mathrm{d}x \tag{9.36}$$

Thus, by definition, we have from Eq. (9.32)

$$\epsilon = \frac{\mathrm{d}^2v}{\mathrm{d}x^2} = 2A_2 + 6A_3x = \{0 \quad 0 \quad 2 \quad 6x\}\{A\} \tag{9.37}$$

or, introducing

$$\{N\} = \{0 \quad 0 \quad 2 \quad 6x\} \tag{9.38}$$

and using Eq. (9.24), we can write Eq. (9.37) in the symbolic form as

$$\epsilon = \{N\}\{A\} = \{N\}[c^{-1}]\{\delta\} \tag{9.39}$$

Then, a comparison between Eqs. (9.7) and (9.39) shows that

$$\{f\} = \{N\}[c^{-1}] \tag{9.40}$$

Substituting this $\{f\}$ into Eq. (9.36) and using the rule of 'transpose multiplication' (see section 9.3), we can show that

$$[K_i] = E_i I_i [c^{-1}]^{\mathrm{T}} \int_0^{h_i} \{N\}^{\mathrm{T}}\{N\}\, \mathrm{d}x [c^{-1}] \tag{9.41}$$

Now, from Eq. (9.38), we can show that

$$\int_0^{h_i} \{N\}^{\mathrm{T}}\{N\}\, \mathrm{d}x = \int_0^{h_i} \begin{bmatrix} 0 & 0 & 0 & 0 \\ 0 & 0 & 0 & 0 \\ 0 & 0 & 4 & 12x \\ 0 & 0 & 12x & 36x^2 \end{bmatrix} \mathrm{d}x = \begin{bmatrix} 0 & 0 & 0 & 0 \\ 0 & 0 & 0 & 0 \\ 0 & 0 & 4h_i & 6h_i^2 \\ 0 & 0 & 6h_i^2 & 12h_i^3 \end{bmatrix}$$

Using this result and $[c^{-1}]$ of Eq. (9.35), we can show by forming the product of Eq. (9.41) that $[K_i]$ obtained in this way is exactly the same as that of Chapter 2.

Distributed external forces

Let us write Eq. (9.32) as

$$U = \{M\}\{A\} \tag{9.42}$$

in which

$$\{M\} = \{1 \quad x \quad x^2 \quad x^3\} \tag{9.43}$$

Elimination of $\{A\}$ between Eqs. (9.24) and (9.42) leads to

$$U = \{M\}[c^{-1}]\{\bar{\delta}\} \tag{9.44}$$

A comparison between Eqs. (9.15) and (9.44) shows that

$$\{b\} = \{M\}[c^{-1}] \tag{9.45}$$

In order to calculate $\{P_e\}$ for the beam element of Fig. 9.4, for example, $\{b\}$ is now substituted into Eq. (9.18); as a result,

$$\{P_e\} = [c^{-1}]^{\mathrm{T}} . \int_{x_1}^{x_2} \{M\}^{\mathrm{T}} . p\, \mathrm{d}x$$

or, using Eqs. (9.35) and (9.43), we can show that the above equation leads to

$$\{P_e\} = \frac{p}{h_i^3} \int_{x_1}^{x_2} \begin{Bmatrix} (h_i^3 - 3h_i x^2 + 2x^3) \\ (h_i^3 x - 2h_i^2 x^2 + h_i x^3) \\ (3h_i x^2 - 2x^3) \\ (h_i x^3 - h_i^2 x^2) \end{Bmatrix} \mathrm{d}x \tag{9.46}$$

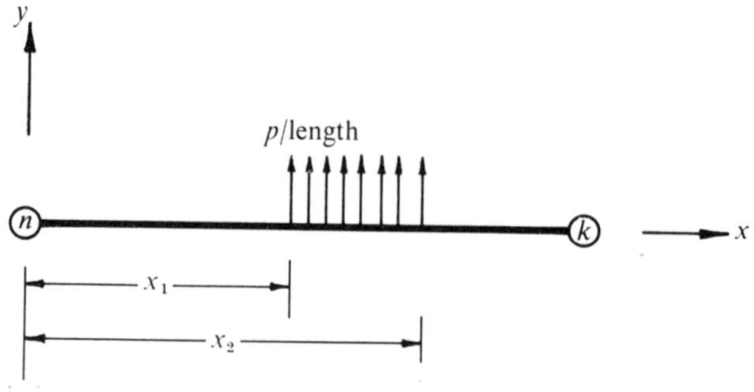

Fig. 9.4 A beam element carrying distributed loading.

$\{P_e\}$ obtained from this equation with $x_1 = 0$ and $x_2 = h_i$ agrees exactly with that of Eq. (5.8).

Concentrated loads

An equivalent load vector for concentrated external loads can also be derived from Eq. (9.18), by imagining the concentrated load to act over a very small area. As an example, we may assume that in the case of Fig. 9.5

$$p = \frac{W}{2e}$$

where e is very small. $\{P_e\}$ for this case can now be calculated from Eq. (9.46), by substituting into it this value of p and setting $x_1 = (L - e)$ and $x_2 = (L + e)$. A typical element of $\{P_e\}$, say the second element, can be calculated as follows:

$$P_e(2) = \frac{W}{2h_i^3 e} \int_{(L-e)}^{(L+e)} (h_i^3 x - 2h_i^2 x^2 + h_i x^3)\, dx$$

As e/L is negligibly small, we may write

$$(1 \pm e/L)^n = 1 \pm ne/L$$

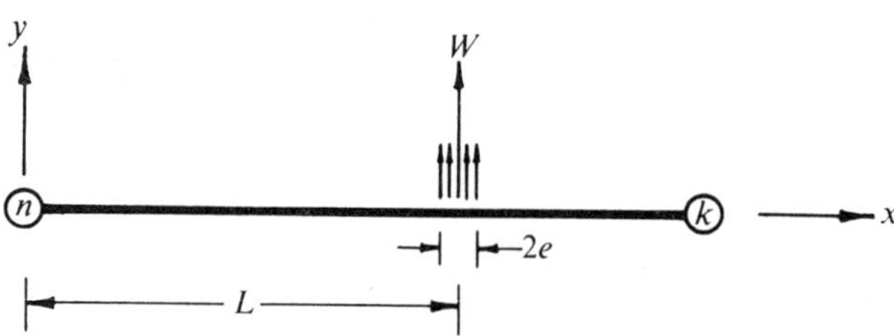

Fig. 9.5 Approximation of the point load W as a distributed load of intensity $p = W/2e$, where e is very small.

Using this in the result of the above integral and simplifying, we can show that

$$P_e(2) = \frac{W}{h_i^3}(h_i^3L - 2h_i^2L^2 + h_iL^3)$$

We can show in this way that for the element of Fig. 9.5,

$$\{P_e\} = \frac{W}{h_i^3}\begin{Bmatrix} (h_i^3 - 3h_iL^2 + 2L^3) \\ (h_i^3L - 2h_i^2L^2 + h_iL^3) \\ (3h_iL^2 - 2L^3) \\ (h_iL^3 - h_i^2L^2) \end{Bmatrix} \tag{9.47}$$

The vector of Eq. (9.47), when used, will obviously eliminate the need to provide a node at each point load. This is particularly useful in dealing with structural members carrying multiple externally applied point loads.

9.8 Summary of equations and observations

The various matrices of a finite element are related as follows:

$$\{\bar{P}\} = [K_i]\{\bar{\delta}\}$$
$$\{U\} = [M]\{A\} = [b]\{\bar{\delta}\}^*$$
$$\{\epsilon\} = [N]\{A\} = [f]\{\bar{\delta}\}^*$$
$$\{A\} = [c^{-1}]\{\bar{\delta}\}$$
$$[f] = [Nc^{-1}]$$
$$[b] = [Mc^{-1}]$$

where
$\{\bar{P}\}$ = vector of extremity forces,
$\{\bar{\delta}\}$ = vector of extremity displacements,
$\{U\}$ = assumed displacement function,
$\{A\}$ = vector of constants defining $\{U\}$, and
$\{\epsilon\}$ = strains within the element.

The various characteristics of the element are given by the following formulae:

Element stiffness matrix (Eq. 9.11)

$$[K_i] = \int_V [f]^T \cdot [d] \cdot [f]\, dv = [c^{-1}]^T \cdot \int_V [N]^T \cdot [d] \cdot [N]\, dv[c^{-1}]$$

* $\{U\}$ and $\{\epsilon\}$ become scalar quantities when they define single components of displacement and strain respectively.

Thermal force vector (Eq. 9.12)

$$\{P_t\} = \int_V [f]^T \cdot \{\sigma\}_t \, dv = [c^{-1}]^T \cdot \int_V [N]^T \cdot \{\sigma\}_t \, dv$$

Equivalent force vector (Eq. 9.18)

$$\{P_e\} = \int_S [b]^T \cdot p \, ds = [c^{-1}]^T \cdot \int_S [M]^T \cdot p \, ds$$

Body force vector (Eq. 9.19)

$$\{P_b\} = \int_V [b]^T \cdot \begin{Bmatrix} \bar{X} \\ \bar{Y} \\ \bar{Z} \end{Bmatrix} dv = [c^{-1}]^T \cdot \int_V [M]^T \cdot \begin{Bmatrix} \bar{X} \\ \bar{Y} \\ \bar{Z} \end{Bmatrix} dv$$

Initial strain force vector (Eq. 9.13)

$$\{P_i\} = \int_V [f]^T \cdot [d] \cdot \{e_i\} \, dv = [c^{-1}]^T \cdot \int_V [N]^T \cdot [d] \cdot \{e_i\} \, dv$$

The vectors $\{P_t\}$, $\{P_e\}$, $\{P_b\}$ and $\{P_i\}$ must be applied as external forces at the nodes to which the element is attached, in addition to all other external forces which these nodes may have. The precise manner of application of these forces is as follows:

Consider, for instance, the application of $\{P_b\}$ to the element of Fig. 11.2. Since this element has 3 extremities and 2 degrees of freedom, its $\{P_b\}$ will be a 6-component vector; let the computed value of $\{P_b\}$ be

$$\{P_b\} = \begin{Bmatrix} F_n \\ Q_n \\ F_k \\ Q_k \\ F_m \\ Q_m \end{Bmatrix}$$

Then, these components will be applied to the element as external forces as shown in Fig. 11.2. More generally, if the degree of freedom of the element is m, then the first m components of $\{P_b\}$ will act as external forces at the node attached to extremity a, the next m components at the node attached to b, and so on.

Observations

We shall make the following observations to conclude this chapter:

(a) The equations of this section show that the various characteristics of a finite element are functions of the three matrices $[M]$, $[c]$ and $[N]$, which can therefore be regarded as the 'primary' matrices of the element. Accordingly, in future discussions on the characteristics of various elements our main objective would be to derive their primary matrices. Furthermore, it is not necessary to write down the $S(i, J, K)$'s of a given element explicitly; instead, it is more convenient and indeed desirable to instruct the computer to identify the $S(i, J, K)$'s from $[K_i]$ automatically.

(b) According to assumption 3 of section 2.1, when $[d]$, ρ or indeed any other property of the body varies from point to point, the average values of these properties within each element will be treated as being constant throughout that element. However, this assumption will normally be overruled if the variation of these properties can be represented by explicit continuous functions, such that it is possible to evaluate the integrals of the formulae given in this section in the closed form. Thus, the beam of Fig. 4.4b, for instance, can be subdivided alternatively into a number of non-uniform elements. For the same number of elements, this will lead to relatively more accurate results.

As an example of this, consider the non-uniform, homogeneous bar element of Fig. 9.6 whose Young's modulus is E. If, according to assumption 3 of section

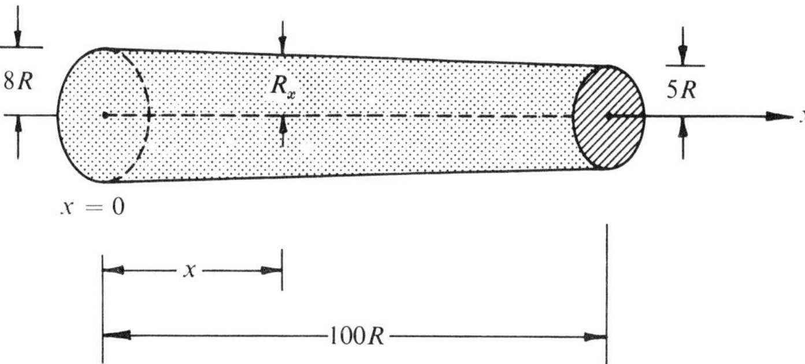

Fig. 9.6 A tapered bar element having a circular cross section.

2.1, this element is treated as a uniform bar element of radius $6.5R$, then from Eq. (9.31) its stiffness matrix will be

$$[K_i] = 0{\cdot}4225\pi RE\begin{bmatrix} 1 & -1 \\ -1 & 1 \end{bmatrix}$$

However, if we treat it as a non-uniform bar whose area of cross-section varies as

$$a(x) = \pi R_x^2 = \pi(8R - 0{\cdot}03x)^2$$

then Eq. (9.30) will give

$$[K_i] = \frac{E}{10^4 R^2}\begin{bmatrix} 1 & -1 \\ -1 & 1 \end{bmatrix}\int_0^{100R} a(x)\,\mathrm{d}x = 0{\cdot}43\pi RE\begin{bmatrix} 1 & -1 \\ -1 & 1 \end{bmatrix}$$

The difference between these results is worth noting.

(c) Consider the uniform bar element shown in Fig. 9.7 in which only axial

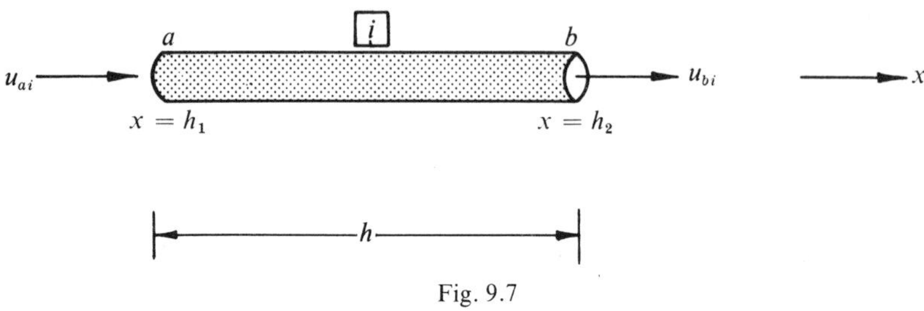

Fig. 9.7

forces and displacements occur. Now, from Eq. (9.20), we have

$$u_{ai} = u(h_1) = A_0 + A_1 h_1$$

and

$$u_{bi} = u(h_2) = A_0 + A_1 h_2$$

We can therefore show that

$$A_0 = u_{ai} - \frac{(u_{bi} - u_{ai})h_1}{h} = u_{bi} - \frac{(u_{bi} - u_{ai})h_2}{h} \tag{9.48}$$

Now, suppose that this element undergoes a 'small' rigid-body displacement (section 4.1) along x by the amount β. Then, since all points in the bar undergo exactly the same rigid-body displacement, we have

$$u_{ai} = u_{bi} = \beta$$

Consequently, Eq. (9.48) reduces to

$$A_0 = \beta$$

indicating that the 'free' constant A_0 in Eq. (9.20) in fact represents the rigid-body displacement of the bar element. For a given finite element, we can demonstrate in this way that in general, a 'free' constant (such as A_0 above), which is included in the assumed expression for U, represents a rigid-body displacement mode of the element.

In all finite elements, a free constant must be included in the assumed expression for each mutually independent component of displacement. Failure to do so would lead to the physically inconsistent situation in which the element is not able to undergo a rigid-body displacement.

References

1. Argyris, J. H., *Energy Theorems and Structural Analysis*, Butterworths Scientific Publications, London, 1960.

2. Bazeley, G. P., *et. al.*, 'Triangular elements in plate bending: conforming and non-conforming solutions', *Proc. conf. Matrix methods in Structural Mechanics*, Wright Patterson Air Force Base, Ohio, 1965.

10 Plate and shell structures

10.1 General

Plates, shells and folded roofs constitute an important category of structures whose analysis by conventional methods is generally difficult. Finite elements, on the other hand, can be used very effectively for their accurate analysis. These structures are normally designed to sustain transverse forces and moments as well as 'in-plane' forces. An 'in-plane' force is defined as a force that lies on the plane of the element (these forces are not a common feature of flat plate structures). Also, observe that the in-plane forces are to a plate what axial forces are to a beam.

By virtue of the assumption of geometrical linearity (section 2.1), the 'bending stiffness' of a plate is not 'coupled'* to its 'in-plane stiffness', just as the bending stiffness of the beam element (Chapter 2) was unaffected by its axial stiffness. Each set of stiffnesses of finite elements to deal with these problems can therefore be derived, as in the case of the beam element, independently of the other.

In this chapter we shall first deal with the 'flat plate structures', defined as flat plates subjected to bending only (no in-plane forces). The problems of 'stretched' or 'compressed' plates, defined as flat plates subjected to both transverse and in-plane forces, will be considered in section 10.8. Finally, in section 10.9 we shall discuss the treatment of shell and folded roof structures.

When dealing with plate and shell problems, the structure will be subdivided into what we shall call 'flat plate' or simply 'plate' elements, which are in fact two-dimensional finite elements of constant thickness. Plate elements of triangular, rectangular and quadrilateral shapes can be used for analysis. The use of the quadrilateral element, however, is not an attractive proposition, as the derivation of its characteristics involves integrals which are difficult to evaluate, even numerically. In any case a quadrilateral element can be looked upon as a combination of two triangular elements. We shall therefore confine the present discussion to triangular and rectangular elements only. The problems posed by the quadrilateral element will be discussed in section 11.3.

As pointed out in section 9.8, the stiffness and other characteristics of a finite element can be derived from its 'primary' matrices $[c]$, $[M]$ and $[N]$. Consequently, our main objective here, as elsewhere, would be to obtain the explicit forms of these matrices. Moreover, as we shall use the computer to perform the various phases of calculation, it is pointless to write down the

* In this context 'coupling' means interaction.

$S(i, J, K)$'s of the plate elements explicitly. Instead, it is more convenient and indeed desirable to instruct the computer to form the $[K_i]$'s, and then identify from them the $S(i, J, K)$'s automatically.

10.2 Stress-strain relationship

For deriving the 'elasticity' matrix $[d]$ for flat plates, we shall make the following assumptions:

(1) The 'mid-plane' (Fig. 10.1) of the unloaded plate is flat and lies on the x–y plane.

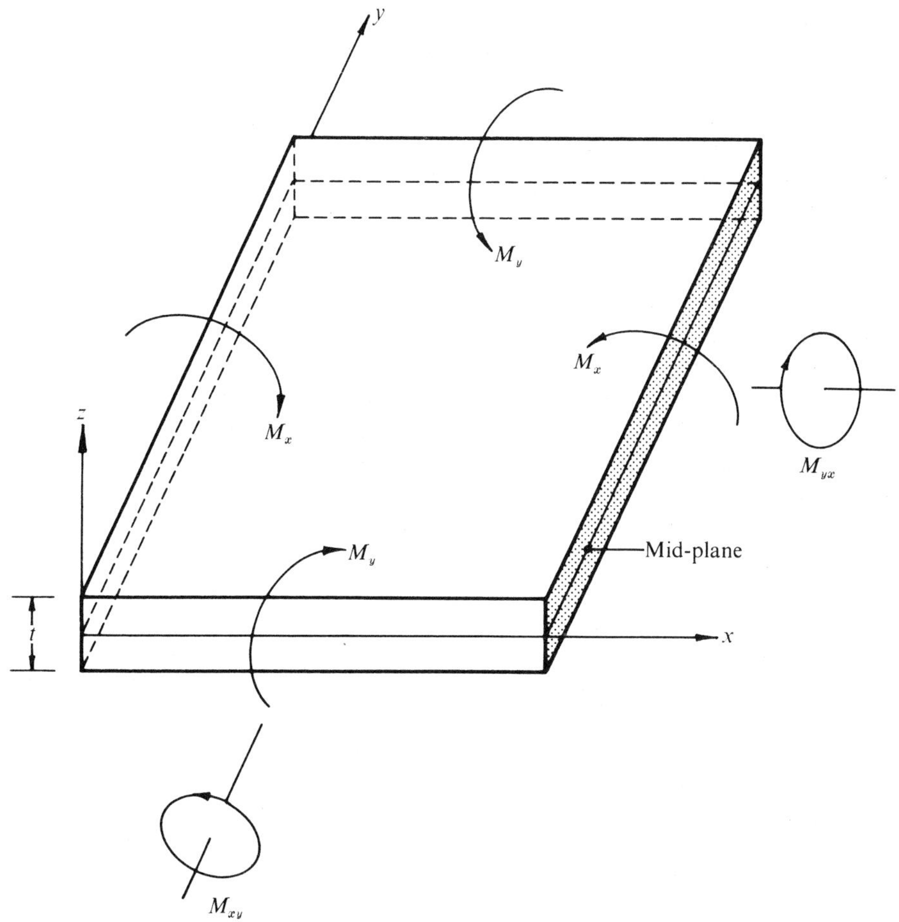

Fig. 10.1 'Stress moments' in a plate.

(2) The thickness of the plate is small compared with its other dimensions and,

(3) the material of the plate is isotropic, homogeneous and linearly elastic (orthotropy will be discussed later).

Then, from the classical Kirchhoff plate theory for small displacements, we can show that the stresses within the plate are related to its transverse deflection $w(x, y)$ along z as[1]:

$$\sigma_{xx} = \frac{-zE}{(1-\nu^2)}\left(\frac{\partial^2 w}{\partial x^2} + \nu\frac{\partial^2 w}{\partial y^2}\right) \tag{10.1a}$$

$$\sigma_{yy} = \frac{-zE}{(1-\nu^2)}\left(\frac{\partial^2 w}{\partial y^2} + \nu\frac{\partial^2 w}{\partial x^2}\right) \tag{10.1b}$$

and

$$\tau_{xy} = \tau_{yx} = -2Gz\frac{\partial^2 w}{\partial x\, \partial y} \tag{10.1c}$$

in which E, G and ν are Young's modulus, modulus of rigidity and Poisson's ratio respectively. We shall now define the generalized 'strains' within the deformed plate as

$$\{\epsilon\} = \left\{\begin{array}{c} -\dfrac{\partial^2 w}{\partial x^2} \\[2ex] -\dfrac{\partial^2 w}{\partial y^2} \\[2ex] -2\dfrac{\partial^2 w}{\partial x\, \partial y} \end{array}\right\} \tag{10.2}$$

Then, since $E = 2(1+\nu)G$, Eqs. (10.1) can be written in the matrix form as

$$\left\{\begin{array}{c} \sigma_{xx} \\ \sigma_{yy} \\ \tau_{xy} \end{array}\right\} = \frac{Ez}{(1-\nu^2)}\begin{bmatrix} 1 & \nu & 0 \\ \nu & 1 & 0 \\ 0 & 0 & \dfrac{1-\nu}{2} \end{bmatrix}\{\epsilon\} \tag{10.3}$$

Eq. (10.3) obviously specifies the stress-strain relationship for the plate. It is interesting to observe that but for the factor z, the elasticity matrix of Eq. (10.3) is identical to that for plane stress (Eq. 8.19). This is to be expected, however, since, as in plane stress, the Kirchhoff theory also assumes

$$\sigma_{zz} = \tau_{zx} = \tau_{zy} = 0$$

at $-t/2 \leqslant z \leqslant t/2$. As a result, the strain e_{zz} does not vanish and is given by Eq. (8.20).

The stresses within the plate are related to what are called its 'stress moments', M_x, M_y, M_{xy} and M_{yx} as follows:

$$M_x = \int \sigma_{xx} z \, dz, \qquad M_y = \int \sigma_{yy} z \, dz \quad \text{and} \quad M_{yx} = M_{xy} = \int \tau_{xy} z \, dz$$

where the integration is performed, in each case, between the limits $-t/2$ and $t/2$. Now, substituting for σ_{xx} from Eq. (10.1a) into the first of the above equations and integrating, we have

$$M_x = -D\left(\frac{\partial^2 w}{\partial x^2} + \nu \frac{\partial^2 w}{\partial y^2}\right) \tag{10.4a}$$

where the quantity
$$D = \frac{Et^3}{12(1 - \nu^2)}$$

is called the 'bending rigidity' of the plate. Similarly, we can show that

$$M_y = -D\left(\frac{\partial^2 w}{\partial y^2} + \nu \frac{\partial^2 w}{\partial x^2}\right) \tag{10.4b}$$

and
$$M_{xy} = M_{yx} = -D(1 - \nu)\frac{\partial^2 w}{\partial x \, \partial y} \tag{10.4c}$$

Physically, the quantities M_x and M_y represent bending moments per unit length while M_{xy} represents twisting moment, also per unit length. Now, if we define the generalized 'stresses' within the plate as

$$\{\sigma\} = \begin{Bmatrix} M_x \\ M_y \\ M_{xy} \end{Bmatrix}$$

then, Eqs. (10.4) can be written in the following compact matrix form:

$$\{\sigma\} = [d]\{\epsilon\} \tag{10.5}$$

Here, $[d]$ is the required elasticity matrix of the plate, given by

$$[d] = D\begin{bmatrix} 1 & \nu & 0 \\ \nu & 1 & 0 \\ 0 & 0 & \dfrac{1 - \nu}{2} \end{bmatrix}$$

We shall use this $[d]$ for analyzing isotropic plates.

10.3 The rectangular 'plate' element

Fig. 10.2 shows that each extremity of this element has three components of displacement, one of transverse translation and two of rotation. The element's degree of freedom is therefore three. Also, as it has four extremities, we can

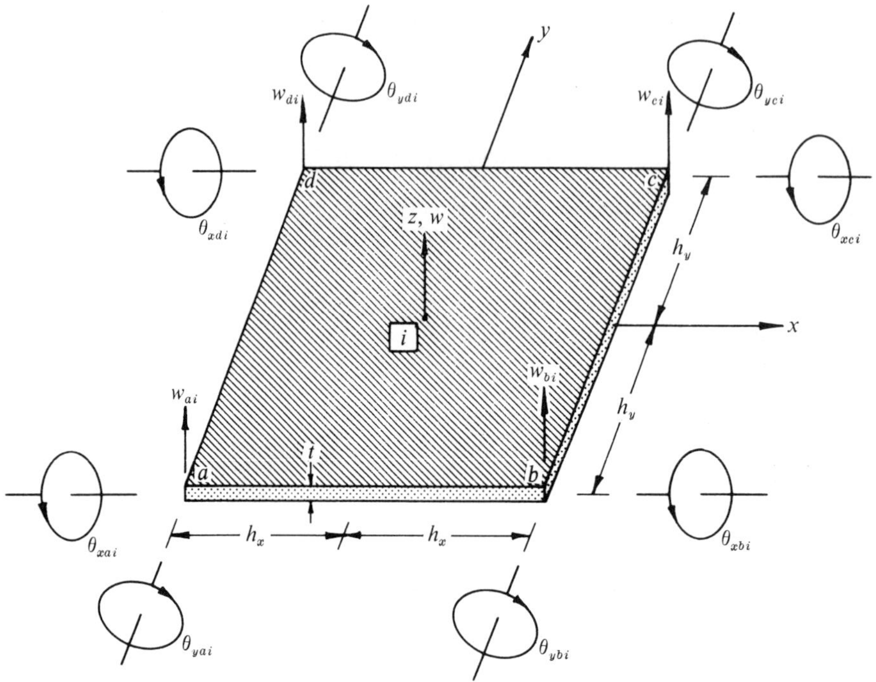

Fig. 10.2 A rectangular 'plate' element in bending. (As usual, θ_{xci} means rotation θ_x at extremity c of element i etc.).

assume its transverse deflection, $w(x, y)$, as a polynomial defined by 12 constants.* In particular, let

$$w(x, y) = A_0 + A_1x + A_2y + A_3x^2 + A_4xy + A_5y^2 + A_6x^3 + A_7x^2y + A_8xy^2 + A_9y^3 + A_{10}x^3y + A_{11}xy^3 \tag{10.6}$$

Then, with
$$\{A\} = \{A_0 \quad A_1 \quad \cdots \quad A_{10} \quad A_{11}\}^{\mathrm{T}}$$

and
$$\{M\} = \{1 \; x \; y \; x^2 \; xy \; y^2 \; x^3 \; x^2y \; xy^2 \; y^3 \; x^3y \; xy^3\} \tag{10.7}$$

we can write Eq. (10.6) in the matrix form as

$$U = w(x, y) = \{M\}\{A\} \tag{10.8}$$

* According to (a) in section 9.5, in this case $n = 4$. Also, since θ_x and θ_y are derived from $w(x, y)$, $m = 2$.

The rotations, $\theta_x(x, y)$ and $\theta_y(x, y)$, are now obtained by observing that

$$\theta_x = \frac{\partial w}{\partial y} \quad \text{and} \quad \theta_y = -\frac{\partial w}{\partial x}$$

Then, differentiating Eq. (10.6) we have

$$\theta_x(x, y) = A_2 + A_4x + 2A_5y + A_7x^2 + 2A_8xy + 3A_9y^2 + A_{10}x^3 + 3A_{11}xy^2 \tag{10.9a}$$

and

$$\theta_y(x, y) = -A_1 - 2A_3x - A_4y - 3A_6x^2 - 2A_7xy - A_8y^2 - 3A_{10}x^2y - A_{11}y^3 \tag{10.9b}$$

As explained in section 9.8(c), the constant A_0 represents the rigid-body translation of the element along z, while A_1 and A_2 represent its rigid-body rotations.

The displacements at extremity j ($j = a, b, c, d$) are now obtained by substituting the extremity coordinates (x_j, y_j) into Eqs. (10.6) and (10.9). Thus, at extremity a ($j = a$) for example, we have from Eq. (10.6)

$$w_{ai} = w(x_a, y_a) = A_0 + A_1x_a + \cdots + A_{11}x_ay_a^3$$

etc. Similarly, for extremity j of element i, we have from Eqs. (10.9)

$$\theta_{xji} = \theta_x(x_j, y_j) = A_2 + A_4x_j + \cdots + 3A_{11}x_jy_j^2$$

and

$$\theta_{yji} = \theta_y(x_j, y_j) = -A_1 - 2A_3x_j - \cdots - A_{11}y_j^3$$

Having thus determined the expressions for all the extremity displacements, we shall now put them in the matrix form as

$$\{\bar{\delta}\} = [c]\{A\}, \tag{10.10}$$

where the vector $\{\bar{\delta}\}$ lists all the extremity displacements as

$$\{\bar{\delta}\} = \{w_{ai}\ \theta_{xai}\ \theta_{yai}\ w_{bi}\ \theta_{xbi}\ \theta_{ybi}\ w_{ci}\ \theta_{xci}\ \theta_{yci}\ w_{di}\ \theta_{xdi}\ \theta_{ydi}\}^{\mathrm{T}} \tag{10.11}$$

and the primary matrix $[c]$ is as given in Table 10.1. Obviously, $[c]$ depends only on the coordinates of the extremities. This is true of all finite elements.

The primary matrix $[N]$ is now obtained from Eq. (10.2): thus, by differentiating the assumed deflection $w(x, y)$ of Eq. (10.6) according to the scheme of Eq. (10.2), we can show that

$$\{\epsilon\} = -\begin{bmatrix} 0 & 0 & 0 & 2 & 0 & 0 & 6x & 2y & 0 & 0 & 6xy & 0 \\ 0 & 0 & 0 & 0 & 0 & 2 & 0 & 0 & 2x & 6y & 0 & 6xy \\ 0 & 0 & 0 & 0 & 2 & 0 & 0 & 4x & 4y & 0 & 6x^2 & 6y^2 \end{bmatrix}\{A\} \tag{10.12}$$

or symbolically

$$\{\epsilon\} = [N]\{A\} \tag{10.13}$$

Table 10.1 The $[c]$ matrix for a rectangular 'plate' element

$[c] =$

	1	2	3	4	5	6	7	8	9	10	11	12
1	1	x_a	y_a	x_a^2	$x_a y_a$	y_a^2	x_a^3	$x_a^2 y_a$	$x_a y_a^2$	y_a^3	$x_a^3 y_a$	$x_a y_a^3$
2	0	0	1	0	x_a	$2y_a$	0	x_a^2	$2x_a y_a$	$3y_a^2$	x_a^3	$3x_a y_a^2$
3	0	-1	0	$-2x_a$	$-y_a$	0	$-3x_a^2$	$-2x_a y_a$	$-y_a^2$	0	$-3x_a^2 y_a$	$-y_a^3$
4	1	x_b	y_b	x_b^2	$x_b y_b$	y_b^2	x_b^3	$x_b^2 y_b$	$x_b y_b^2$	y_b^3	$x_b^3 y_b$	$x_b y_b^3$
5	0	0	1	0	x_b	$2y_b$	0	x_b^2	$2x_b y_b$	$3y_b^2$	x_b^3	$3x_b y_b^2$
6	0	-1	0	$-2x_b$	$-y_b$	0	$-3x_b^2$	$-2x_b y_b$	$-y_b^2$	0	$-3x_b^2 y_b$	$-y_b^3$
7	1	x_c	y_c	x_c^2	$x_c y_c$	y_c^2	x_c^3	$x_c^2 y_c$	$x_c y_c^2$	y_c^3	$x_c^3 y_c$	$x_c y_c^3$
8	0	0	1	0	x_c	$2y_c$	0	x_c^2	$2x_c y_c$	$3y_c^2$	x_c^3	$3x_c y_c^2$
9	0	-1	0	$-2x_c$	$-y_c$	0	$-3x_c^2$	$-2x_c y_c$	$-y_c^2$	0	$-3x_c^2 y_c$	$-y_c^3$
10	1	x_d	y_d	x_d^2	$x_d y_d$	y_d^2	x_d^3	$x_d^2 y_a$	$x_d y_d^2$	y_d^3	$x_d^3 y_d$	$x_d y_d^3$
11	0	0	1	0	x_d	$2y_d$	0	x_d^2	$2x_d y_d$	$3y_d^2$	x_d^3	$3x_d y_d^2$
12	0	-1	0	$-2x_d$	$-y_d$	0	$-3x_d^2$	$-2x_d y_d$	$-y_d^2$	0	$-3x_d^2 y_d$	$-y_d^3$

since by definition (section 9.8), the rectangular matrix of Eq. (10.12) is the required $[N]$ matrix (in this case it incorporates the negative sign).

Having thus determined the three primary matrices, we can now derive the various element characteristics from the formulae of section 9.8; for instance, with the $[d]$ matrix defined by Eq. (10.5), we may write

$$[K_i] = [c^{-1}]^{\mathrm{T}}[\bar{K}][c^{-1}] \tag{10.14}$$

where

$$[\bar{K}] = \int_S [N]^{\mathrm{T}}[d][N]\,\mathrm{d}x\,\mathrm{d}y \tag{10.15}$$

in which the integration is performed over the entire area of the element. The explicit form of $[\bar{K}]$ is given in Table 10.2 (note that here the centroid of the element has been taken as the origin of coordinates, as in Fig. 10.2). $[K_i]$ can now be obtained by forming the matrix product of Eq. (10.14); however, as some elements on the leading diagonal of $[c]$ (Table 10.1) are zero, we are unable to form $[c]^{-1}$ by direct inversion. This difficulty is easily overcome by the simple procedures given in section 15.3. Having thus computed the individual $[K_i]$'s of all elements into which the plate is subdivided, its overall stiffness matrix, $[K]$, is then assembled by the methods of Chapter 3. However, in view of the large sizes of matrices involved here, we shall leave this process entirely to the computer; relevant programs for this will be discussed in section 15.5. The problem can then be solved in terms of Eq. (3.9), having accounted for the prescribed boundary conditions. If the plate contains a total of N nodes, then the listings of $\{P\}$ and $\{\delta\}$ will be as follows:

$$\{P\} = \{H_1\ M_{x1}\ M_{y1}\quad H_2\ M_{x2}\ M_{y2}\quad \cdots\quad H_N\ M_{xN}\ M_{yN}\}^{\mathrm{T}} \tag{10.16}$$

and

$$\{\delta\} = \{w_1\ \theta_{x1}\ \theta_{y1}\quad w_2\ \theta_{x2}\ \theta_{y2}\quad \cdots\quad w_N\ \theta_{xN}\ \theta_{yN}\}^{\mathrm{T}} \tag{10.17}$$

Table 10.2 The $[\bar{K}]$ matrix for the rectangular element; to be multiplied by $(D \times A)$: $D = \dfrac{Et^3}{12(1 - \nu^2)}$, $A = 4h_x h_y$; $q = \dfrac{(1 - \nu)}{2}$. (Centroid of element taken as the origin of coordinates)

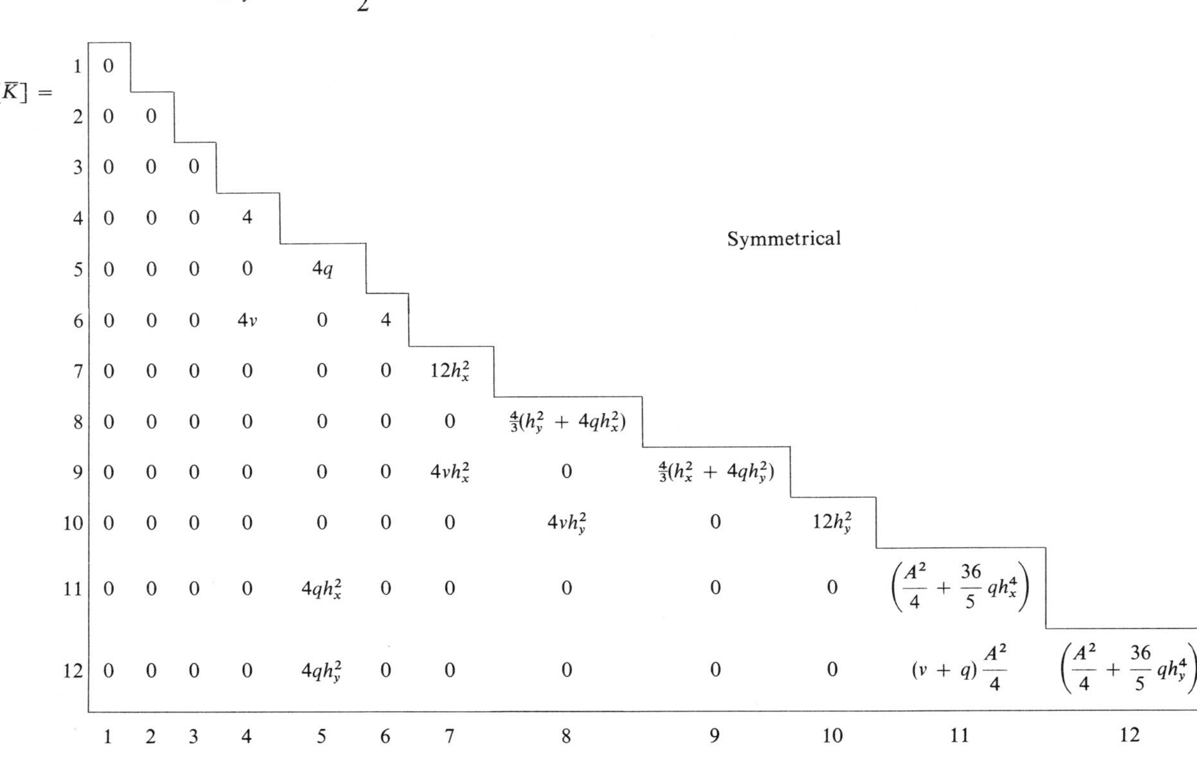

$[\bar{K}] =$

	1	2	3	4	5	6	7	8	9	10	11	12
1	0								Symmetrical			
2	0	0										
3	0	0	0									
4	0	0	0	4								
5	0	0	0	0	$4q$							
6	0	0	0	4ν	0	4						
7	0	0	0	0	0	0	$12h_x^2$					
8	0	0	0	0	0	0	0	$\frac{4}{3}(h_y^2 + 4qh_x^2)$				
9	0	0	0	0	0	0	$4\nu h_x^2$	0	$\frac{4}{3}(h_x^2 + 4qh_y^2)$			
10	0	0	0	0	0	0	0	$4\nu h_y^2$	0	$12h_y^2$		
11	0	0	0	0	$4qh_x^2$	0	0	0	0	0	$\left(\frac{A^2}{4} + \frac{36}{5}qh_x^4\right)$	
12	0	0	0	0	$4qh_y^2$	0	0	0	0	0	$(\nu + q)\frac{A^2}{4}$	$\left(\frac{A^2}{4} + \frac{36}{5}qh_y^4\right)$
	1	2	3	4	5	6	7	8	9	10	11	12

In Fig. 10.2, the arrows indicate the positive directions of various displacements and their respective associated forces. Notice that for node k (= 1, 2, . . . N),

$$\begin{aligned} H_k &= \text{transverse (along } z\text{) external force at } k, \\ w_k &= \text{transverse deflection of } k, \\ M_{xk}, M_{yk} &= \text{externally applied moments at } k \text{ and} \\ \theta_{xk}, \theta_{yk} &= \text{rotations at } k. \end{aligned}$$

When thermal and initial strain forces are present, these must be included in the vector $\{P\}$ in the manner explained in section 9.8.

Once the nodal displacements, $\{\delta\}$, have been computed, the stress-moments within individual elements can be obtained from Eq. (10.5); for this the general expression, including thermal and initial strain forces, when present, is

$$\begin{Bmatrix} M_x \\ M_y \\ M_{xy} \end{Bmatrix} = [d][f]\{\bar{\Delta}\} - \frac{E\alpha T t^3}{12(1-\nu)} \begin{Bmatrix} 1 \\ 1 \\ 0 \end{Bmatrix} - [d] \begin{Bmatrix} e_{ixx} \\ e_{iyy} \\ e_{ixy} \end{Bmatrix} \tag{10.18}$$

where

$$\{\bar{\Delta}\} = \{w_n \;\; \theta_{xn} \;\; \theta_{yn} \quad w_k \;\; \theta_{xk} \;\; \theta_{yk} \quad w_m \;\; \theta_{xm} \;\; \theta_{ym} \quad w_j \;\; \theta_{xj} \;\; \theta_{yj}\}^{\mathrm{T}}$$

and, by definition (Chapter 9),

$$[f] = [Nc^{-1}]$$

When using Eq. (10.18), it is important to observe that the nodes n, k, m and j are assumed to be attached to the extremities a, b, c and d respectively, as shown in Fig. 10.3.

Equation (10.12) shows that $[N]$, and hence the stress-moments of Eq. (10.18) are functions of position; thus, their values at a given point within the element will depend on the coordinates of that point. A mathematical consequence of the dependence of $[N]$ on position is that the stresses σ_{xx}, σ_{yy} and τ_{xy} (Eqs. 10.1), in general, violate the stress equilibrium conditions expressed by Eqs. (8.21).

Fig. 10.3 Assumed attachment of nodes to the extremities of a rectangular element.

Interface continuity

As we have seen in Chapter 3, the continuity of slopes and displacements at the nodes is enforced by the assembly procedure of the overall stiffness matrix. Further, as remarked in section 9.5, the assumed displacement function should be such that the displacements and slopes are also continuous across the interfaces between adjacent two- and three-dimensional elements; lack of continuity will, in general, result in a multiplicity of these quantities at the interfaces.

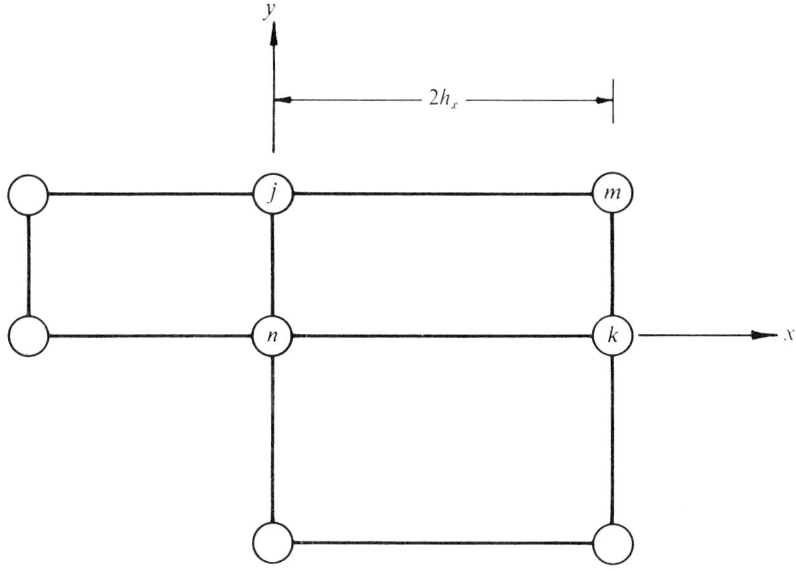

Fig. 10.4 Typical interfaces between adjacent rectangular plate elements.

Along the typical interface $n - k$ ($y = 0$) between two adjacent rectangular elements, shown in Fig. 10.4, we find from Eqs. (10.6) and (10.9b) that $w(x, 0)$ and $\theta_y(x, 0)$ are given by

$$w(x, 0) = A_0 + A_1x + A_3x^2 + A_6x^3 \tag{10.19a}$$

and

$$\theta_y(x, 0) = -A_1 - 2A_3x - 3A_6x^2 \tag{10.19b}$$

Then, from Fig. 10.4 it follows that

$$w_n = w(0, 0)$$
$$\theta_{yn} = \theta_y(0, 0)$$
$$w_k = w(2h_x, 0)$$

and

$$\theta_{yk} = \theta_y(2h_x, 0)$$

Substituting these into Eqs. (10.19) and solving, we can show that

$$A_0 = w_n$$

$$A_1 = -\theta_{yn}$$

$$A_3 = \frac{1}{2h_x}(2\theta_{yn} + \theta_{yk}) + \frac{3}{4h_x^2}(-w_n + w_k)$$

and

$$A_6 = -\frac{1}{4h_x^2}(\theta_{yn} + \theta_{yk}) - \frac{1}{4h_x^3}(-w_n + w_k)$$

Clearly, these constants depend only on the values of $w(x, y)$ and $\theta_y(x, y)$ at the nodes n and k. Thus, it follows from Eqs. (10.19) that the variations of $w(x, y)$ and $\theta_y(x, y)$ along the interface $n - k$ depend only on the values of $w(x, y)$ and $\theta_y(x, y)$ at the nodes n and k. We conclude therefore that $w(x, y)$ and $\theta_y(x, y)$ are continuous across the interface $n - k$.

Equation (10.9a) gives $\theta_x(x, y)$ along $n - k$ as

$$\theta_x(x, 0) = A_2 + A_4 x + A_7 x^2 + A_{10} x^3 \tag{10.20}$$

Clearly, Eqs. (10.19) and (10.20) have no constants in common. Moreover, the only two conditions, namely

$$\theta_{xn} = \theta_x(0, 0)$$

and

$$\theta_{xk} = \theta_x(2h_x, 0)$$

are not sufficient to determine the explicit form of $\theta_x(x, 0)$ from Eq. (10.20). Mathematically, this indeterminacy indicates that $\theta_x(x, y)$ is discontinuous across $n - k$.

Following exactly the same procedure we can also demonstrate that while $w(x, y)$ and $\theta_x(x, y)$ are continuous across interfaces parallel to the y axis (e.g., $n - j$ in Fig. 10.4), $\theta_y(x, y)$ is not. Thus, to summarize: $w(x, y)$ is continuous across all interfaces between adjacent rectangular elements while, $\theta_x(x, y)$ and $\theta_y(x, y)$ are continuous only across interfaces parallel to the y and x axes respectively. These observations also apply to the triangular plate element which follows in the next section.

10.4 The triangular 'plate' element

This element (Fig. 10.5) obviously has 3 extremities, i.e., $n = 3$; also, as in the rectangular element, $m = 2$. Consequently, according to section 9.5(a), its transverse deflection $w(x, y)$ will be assumed as a polynomial defined by 9 constants as follows:

$$w(x, y) = A_0 + A_1 x + A_2 y + A_3 x^2 + A_4 xy + A_5 y^2 + A_6 x^3 + A_7(xy^2 + x^2 y) + A_8 y^3 {}^* \tag{10.21}$$

* This particular function has been chosen for its symmetry in preference to other functions.

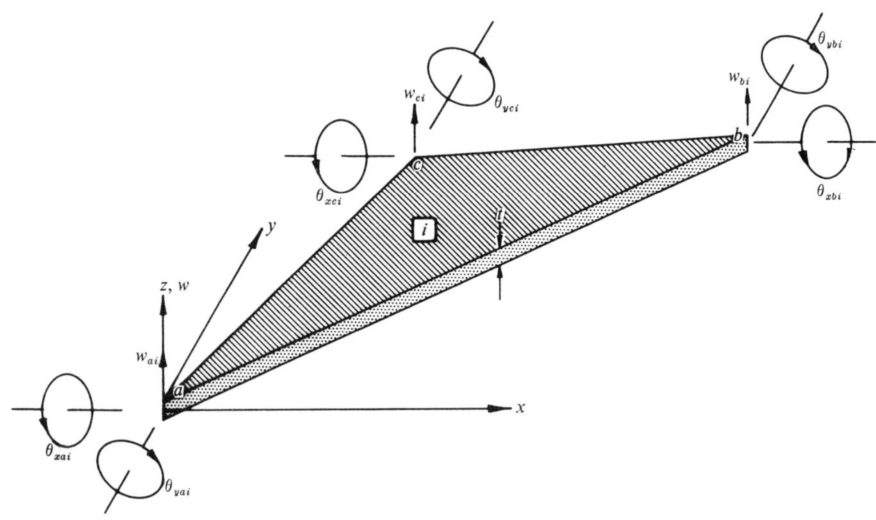

Fig. 10.5 Triangular 'plate' element in bending; the 'middle plane' of the element lies on the x-y plane.

Then, with $\{A\} = \{A_0 \; A_1 \; A_2 \; \cdots \; A_8\}^{\mathrm{T}}$ and

$$\{M\} = \{1 \;\; x \;\; y \;\; x^2 \;\; xy \;\; y^2 \;\; x^3 \;\; (xy^2 + x^2y) \;\; y^3\} \tag{10.21a}$$

we can write Eq. (10.21) in the compact matrix form of Eq. (10.8). By differentiating $w(x, y)$, we have

$$\theta_x(x, y) = \frac{\partial w}{\partial y} = A_2 + A_4x + 2A_5y + A_7(2xy + x^2) + 3A_8y^2 \tag{10.22a}$$

and

$$\theta_y(x, y) = -\frac{\partial w}{\partial x} = -A_1 - 2A_3x - A_4y - 3A_6x^2 - A_7(y^2 + 2xy) \tag{10.22b}$$

As before, the extremity coordinates are substituted into Eqs. (10.21) and (10.22) to give the deflections and rotations at the extremities; thus, at extremity b (coordinates x_b, y_b), for instance,

$$\theta_{xbi} = \theta_x(x_b, y_b) = A_2 + A_4x_b + 2A_5y_b + A_7(2x_by_b + x_b^2) + 3A_8y_b^2$$

etc. Clearly, this process leads to nine such equations, which can be expressed in the matrix form of Eq. (10.10) where now

$$\{\delta\} = \{w_{ai} \;\; \theta_{xai} \;\; \theta_{yai} \quad w_{bi} \;\; \theta_{xbi} \;\; \theta_{ybi} \quad w_{ci} \;\; \theta_{xci} \;\; \theta_{yci}\}^{\mathrm{T}} \tag{10.23}$$

and the $[c]$ matrix as given in Table 10.3.

Table 10.3 The [c] matrix for a triangular plate element

$[c] =$	1	2	3	4	5	6	7	8	9
1	1	x_a	y_a	x_a^2	$x_a y_a$	y_a^2	x_a^3	$x_a y_a^2 + x_a^2 y_a$	y_a^3
2	0	0	1	0	x_a	$2y_a$	0	$2x_a y_a + x_a^2$	$3y_a^2$
3	0	-1	0	$-2x_a$	$-y_a$	0	$-3x_a^2$	$-(y_a^2 + 2x_a y_a)$	0
4	1	x_b	y_b	x_b^2	$x_b y_b$	y_b^2	x_b^3	$x_b y_b^2 + x_b^2 y_b$	y_b^3
5	0	0	1	0	x_b	$2y_b$	0	$2x_b y_b + x_b^2$	$3y_b^2$
6	0	-1	0	$-2x_b$	$-y_b$	0	$-3x_b^2$	$-(y_b^2 + 2x_b y_b)$	0
7	1	x_c	y_c	x_c^2	$x_c y_c$	y_c^2	x_c^3	$x_2 y_c^2 + x_c^2 y_c$	y_c^3
8	0	0	1	0	x_c	$2y_c$	0	$2x_c y_c + x_c^2$	$3y_c^2$
9	0	-1	0	$-2x_c$	$-y_c$	0	$-3x_c^2$	$-(y_c^2 + 2x_c y_c)$	0

As before, the [N] matrix is now obtained by differentiating Eq. (10.21) according to the scheme of Eq. (10.2): it is left to the reader to verify that in this case

$$[N] = -\begin{bmatrix} 0 & 0 & 0 & 2 & 0 & 0 & 6x & 2y & 0 \\ 0 & 0 & 0 & 0 & 0 & 2 & 0 & 2x & 6y \\ 0 & 0 & 0 & 0 & 2 & 0 & 0 & 4(x+y) & 0 \end{bmatrix} \tag{10.24}$$

We are now in a position to derive the various characteristics of the element from the formulae of section 9.8; in this case

$$[\bar{K}] = \int_\Delta [N]^\mathrm{T}[d][N]\,\mathrm{d}x\,\mathrm{d}y \tag{10.25}$$

The explicit form of the product [N^T d N] is given in Table 10.4. [K_i] for the element can now be obtained from Eq. (10.14) and (10.25), together with the [c] matrix of Table 10.3.* [Note that $\int_\Delta \mathrm{d}x\,\mathrm{d}y$ denotes integration over a triangular area.]

The integral of Eq. (10.25), as indeed all integrals to be evaluated for deriving the characteristics of the triangular element, must be taken over the entire area of the triangle. As seen from Appendix 3, the results of such integration assume a simple form when the centroid of the triangle is taken as the origin of co-ordinates; as an example, consider the term on the 8-th row and 6-th column (Table 10.4) of [$\bar{K}$], which is

$$\bar{K}(8, 6) = 4D\int_\Delta (x + \nu y)\,\mathrm{d}x\,\mathrm{d}y$$

From Eqs. (A3.8b and c), it follows at once that

$$\bar{K}(8, 6) = 4DI_{10} + 4D\nu I_{01} = 0 + 0 = 0$$

* Here, as in the rectangular element, [c] cannot be directly inverted as it has zeros on its leading diagonal; the methods of section 15.3 will have to be used for this.

Table 10.4 The $[\bar{K}]$ matrix for the triangular element

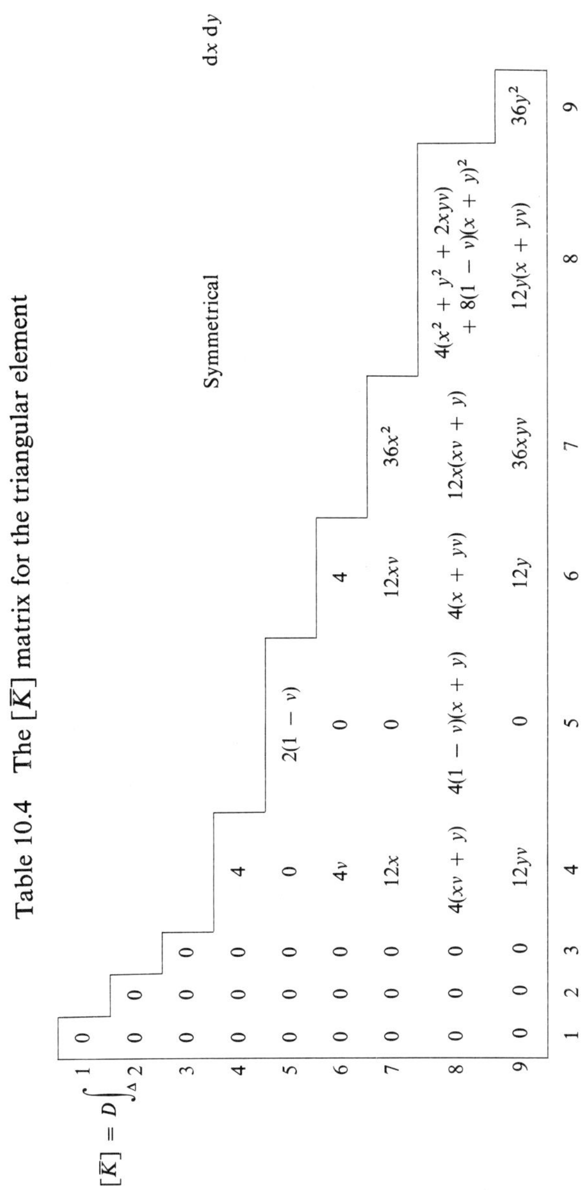

$[\bar{K}] = D \int_{\Delta}$

	1	2	3	4	5	6	7	8	9
1	0								
2	0	0				Symmetrical			
3	0	0	0						
4	0	0	0	4					
5	0	0	0	0	$2(1-\nu)$				
6	0	0	0	4ν	0	4			
7	0	0	0	$12x$	0	$12x\nu$	$36x^2$		
8	0	0	0	$4(x\nu+y)$	$4(1-\nu)(x+y)$	$4(x+y\nu)$	$12x(x\nu+y)$	$4(x^2+y^2+2xy\nu) + 8(1-\nu)(x+y)^2$	
9	0	0	0	$12y\nu$	0	$12y$	$36xy\nu$	$12y(x+y\nu)$	$36y^2$

$\mathrm{d}x\,\mathrm{d}y$

Similarly, with Δ = area of the triangular element,

$$\bar{K}(7,7) = 36D\int_{\Delta} x^2 \, dx \, dy = 36DI_{20}$$
$$= 3D\Delta(x_a^2 + x_b^2 + x_c^2)$$

etc.

The $[K]$ matrix is now assembled in the usual way and adjusted to satisfy the prescribed boundary conditions. The problem may then be solved in terms of Eq. (3.9), with the listings of $\{P\}$ and $\{\delta\}$ given by Eqs. (10.16) and (10.17) respectively. The 'stresses' within individual elements are computed from Eq. (10.18) with

$$\{\bar{\Delta}\} = \{w_n \; \theta_{xn} \; \theta_{yn} \quad w_k \; \theta_{xk} \; \theta_{yk} \quad w_m \; \theta_{xm} \; \theta_{ym}\}^{\mathrm{T}}$$

where the nodes n, k and m are assumed to be attached to the extremities a, b and c respectively.

10.5 The equivalent force vector

Let p denote the uniform intensity of transverse distributed loading on the element. Then, from section 9.8, the equivalent force vector for the element becomes

$$\{P_e\} = p\cdot[c^{-1}]^{\mathrm{T}}\int\{M\}^{\mathrm{T}} \, dx \, dy \qquad (10.26)$$

where the integration has to be performed over the area on which p is active. To illustrate this we consider the triangular element as an example; if its centroid is taken as origin and if p acts over its entire area, then we can show from Eq. (10.21a) that

$$\int_{\Delta}\{M\}^{\mathrm{T}} \, dx \, dy = \{I_{00} \; 0 \; 0 \; I_{20} \; 0 \; I_{02} \; 0 \; 0 \; 0\}^{\mathrm{T}}$$

where I_{00}, I_{20} and I_{02} are given by Eqs. (A3.8a, d and e). We can now find the explicit expression for $\{P_e\}$, if we can derive the explicit form of $[c^{-1}]$. Derivation of the latter for triangular elements is difficult and is not available as of now. Alternatively, $[c^{-1}]$ is easily formed numerically. It then becomes a simple matter to compute $\{P_e\}$ from Eq. (10.26); indeed, this process can be left entirely to the computer.

For rectangular elements, on the other hand, the explicit form of $[c^{-1}]$ is

available[2]; using this and $\{M\}$ of Eq. (10.7), we can show that Eq. (10.26) gives

$$\{P_e\} = \begin{Bmatrix} H_n \\ M_{xn} \\ M_{yn} \\ H_k \\ M_{xk} \\ M_{yk} \\ H_m \\ M_{xm} \\ M_{ym} \\ H_j \\ M_{xj} \\ M_{yj} \end{Bmatrix} = 4ph_xh_y \begin{Bmatrix} \frac{1}{4} \\ -\frac{h_y}{12} \\ \frac{h_x}{12} \\ \frac{1}{4} \\ -\frac{h_y}{12} \\ -\frac{h_x}{12} \\ \frac{1}{4} \\ \frac{h_y}{12} \\ -\frac{h_x}{12} \\ \frac{1}{4} \\ \frac{h_y}{12} \\ \frac{h_x}{12} \end{Bmatrix}$$

where p is assumed to be active over the entire element. The components of $\{P_e\}$ must be applied as external nodal forces, in addition to other external forces. When applying $\{P_e\}$, observe that the nodes n, k, m and j are assumed to be attached to the extremities a, b, c and d, as shown in Fig. 10.3.

Following the method of section 9.7, $\{P_e\}$ for point loads can also be derived, if required, for both rectangular and triangular elements although the calculations will now be relatively more complex.

10.6 An example

As a typical application of finite elements to flat plate structures, let us now solve the problem of Fig. 10.6a, in which a square uniform plate clamped on all

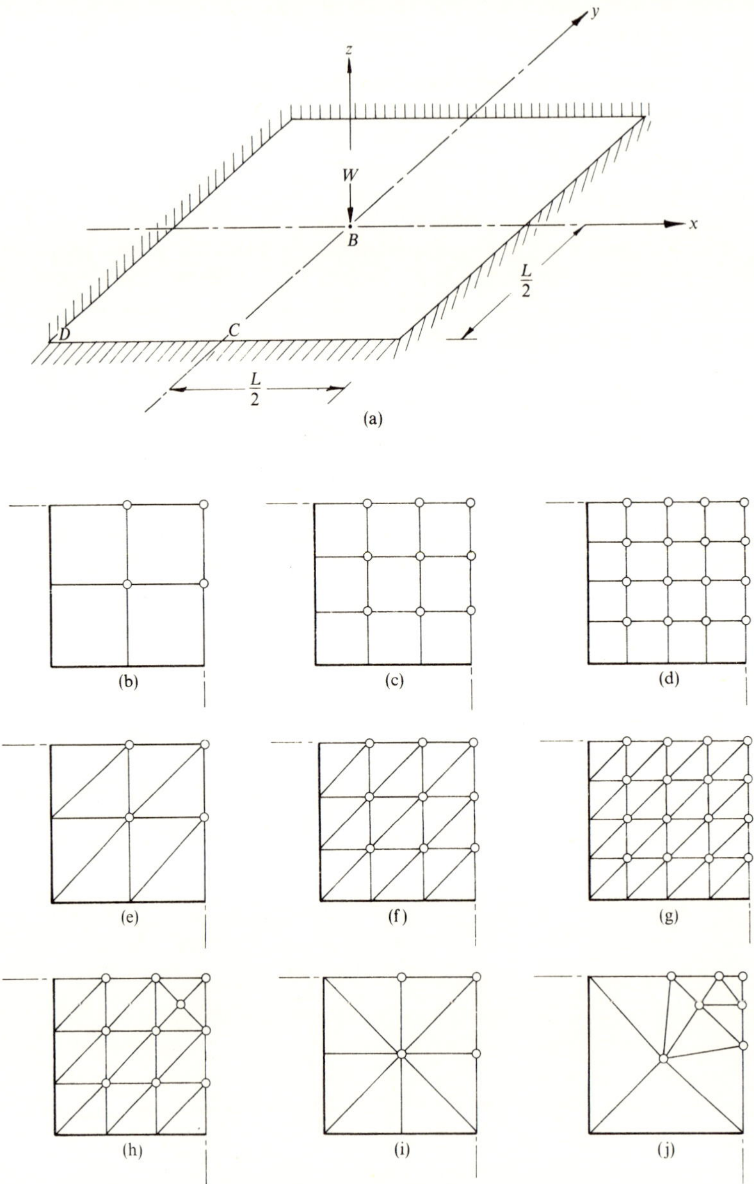

Fig. 10.6 (a) A square, uniform isotropic plate clamped on all sides and subjected to a central point load *W*, acting vertically downwards. (b)–(d) Models of quadrant *ABCD* with square elements. (e)–(j) Models of quadrant *ABCD* with triangular elements.

sides is subjected to a central transverse point load W. Due to the symmetry of this plate, it is sufficient to consider any quadrant of it. Fig. 10.6b–j shows some finite element models of the quadrant $ABCD$*. The boundary conditions for this quadrant are

(a) w, θ_x and θ_y vanish along the clamped sides DA and DC, which need not therefore be represented by nodes

(b) $\theta_y = 0$ on BC and

(c) $\theta_x = 0$ on AB

(b) and (c) arise from the symmetrical deformations of the plate, and are easily dealt with by the methods of section 5.1. (Note that if for some reason we solved this problem by considering the entire plate rather than just one quadrant of it, then conditions (b) and (c) would not arise.)

A simple computer program for solving flat plate problems will be given in section 15.5, where the various steps leading to the solution of this problem will also be outlined. Table 10.5 gives the computed central transverse deflection of the plate for various subdivisions shown in Fig. 10.6.

Table 10.5 shows that when square elements are used, results converge to their exact value with increasingly finer subdivision. In general, this will be found to be the case with rectangular and square elements. Results due to the triangular element, on the other hand, are seen to be relatively less accurate. Also, unlike the square or rectangular elements, it is unwise to try to generalise as to the results to be obtained by using the triangular element. However, this example with the triangular element underlines the observation made in section 9.6, namely that a number of subdivision schemes ought to be tried out to determine how the solution behaves.

Fig. 10.7 shows the computed variation of M_x along AB. The dotted lines

Table 10.5 Computed central transverse deflection of the plate of Fig. 10.6a. $\nu = 0{\cdot}25$

Square elements		Triangular elements	
Model	w*	Model	w*
(b)	0·068747	(e)	0·082682
(c)	0·066345	(f)	0·079558
(d)	0·065186	(g)	0·076543
Exact[1]	0·063000	(h)	0·081458
		(i)	0·063499
		(j)	0·065121

* Multiplier $= -WL^2/Et^3$

* This problem permits us to make the local axes of individual elements parallel to each other. See section 9.5.

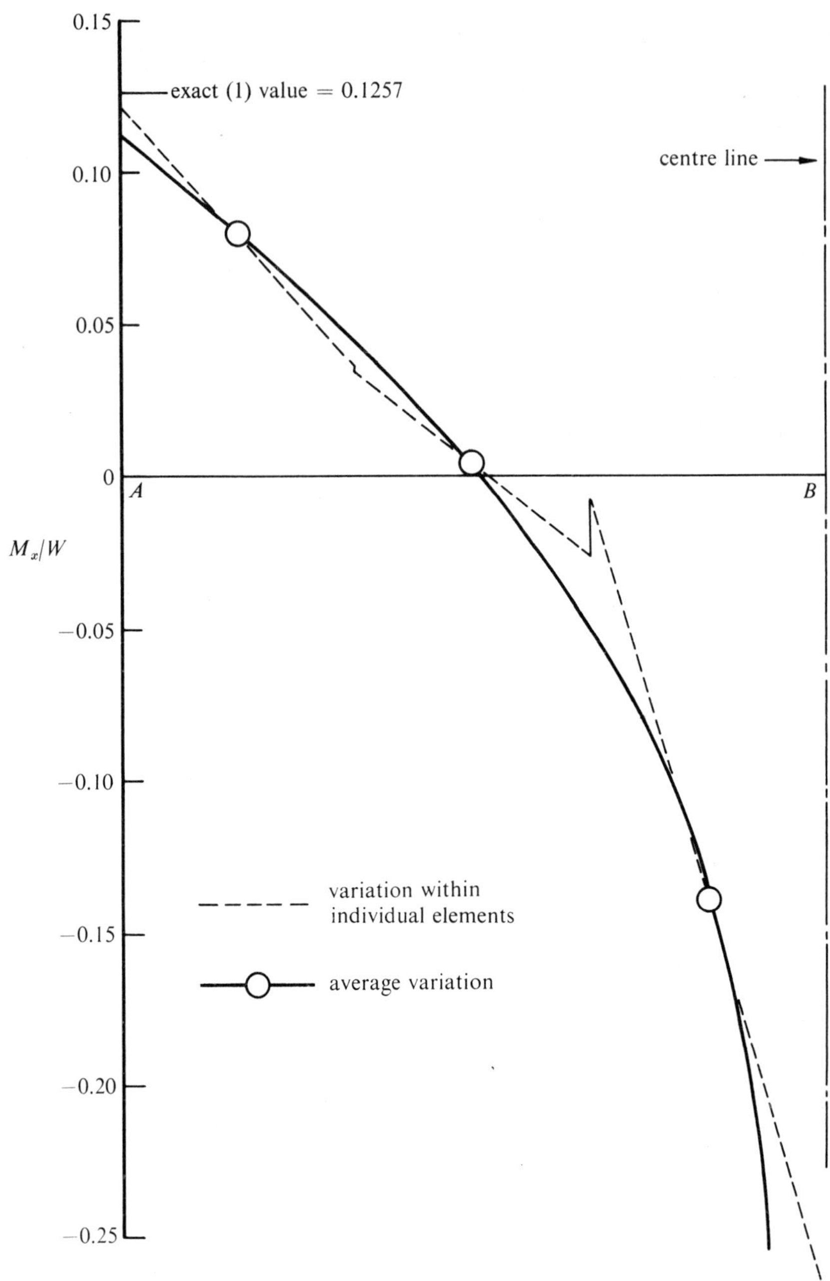

Fig. 10.7 Computed variation of M_x along *AB* of the plate of Fig. 10.6(a), obtained by using model (c).

show that M_x varies linearly within individual elements and that it is discontinuous at common interfaces between adjacent elements. The latter may be attributed to the discontinuity of slopes discussed in section 10.3. However, as the solid line in this figure shows, a smooth and accurate variation of M_x (and of stress moments in general) emerges when its mean value at each element is considered.

Computation of stress moments using triangular elements also leads us to these conclusions, although the interface discontinuities will now be found to be relatively more accentuated.

10.7 Orthotropic plates

So far in our discussion the plate has been assumed to be isotropic, i.e., its elastic properties were the same in all directions. There are instances, however, in which plates exhibit different elastic properties along perpendicular directions; such plates are said to be 'orthotropic'. If the reference axes are chosen such that they coincide with the axes of orthotropy, and assuming that the plate lies on the $x - y$ plane, then, according to Timoshenko and Woinowsky-Krieger,[1] the stresses and strains within it will be related as follows:

$$\begin{aligned} \sigma_{xx} &= E'_x e_{xx} + E'' e_{yy} \\ \sigma_{yy} &= E'' e_{xx} + E'_y e_{yy} \\ \tau_{xy} &= G e_{xy} \end{aligned}$$

Clearly, the behaviour of an orthotropic plate is determined by the four elastic constants, E'_x, E'_y, E'' and G. It can now be demonstrated[1] that the elasticity matrix of the plate becomes

$$[d] = \begin{bmatrix} D_x & D_1 & 0 \\ D_1 & D_y & 0 \\ 0 & 0 & D_{xy} \end{bmatrix} \tag{10.27}$$

in which

$$D_x = E'_x t^3/12, \qquad D_y = E'_y t^3/12, \qquad D_1 = E'' t^3/12 \quad \text{and} \quad D_{xy} = G t^3/12$$

Thus, when analyzing orthotropic plates by finite elements, the $[d]$ matrix of Eq. (10.27) will be used in place of that of Eq. (10.5).

10.8 Analysis of 'stretched' or 'compressed' plates

So far in this chapter we were concerned with flat plates subjected to transverse forces only. Now consider the flat plate of Fig. 10.8 which is acted upon by both transverse and in-plane forces. 'Stretched' or 'compressed' plates of this type can also be analyzed by using the flat plate elements of sections 10.3 and 10.4,

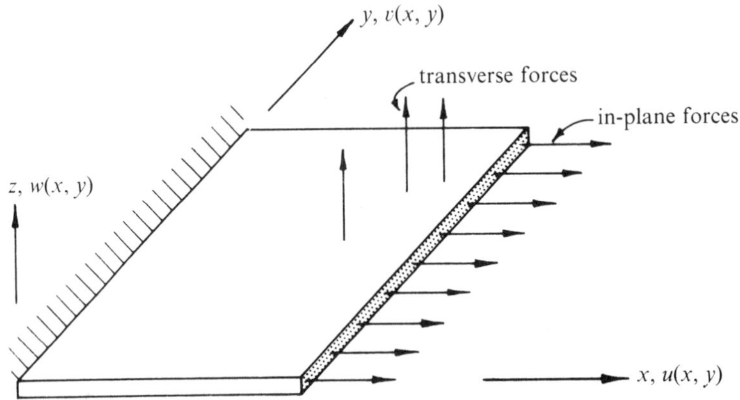

Fig. 10.8 A flat plate subjected to both transverse and in-plane forces.

provided that we modify their characteristics by including the in-plane aspect as well. For future reference, flat plate elements of this type which cater for both transverse and in-plane aspects, will be called 'stretched plate elements'.

Stress-strain relationship for a stretched plate

In addition to the stress-moments of Eq. (10.4), a stretched or compressed plate is also subject to the so called 'stress resultants' N_x, N_y and N_{xy}, as shown in Fig. 10.9. Physically, they represent forces per unit length and in isotropic plates, they are related to the in-plane displacements $u(x, y)$ and $v(x, y)$* as follows[1]:

$$N_x = \psi \frac{\partial u}{\partial x} + \psi \nu \frac{\partial v}{\partial y} \tag{10.28a}$$

$$N_y = \psi \nu \frac{\partial u}{\partial x} + \psi \frac{\partial v}{\partial y} \tag{10.28b}$$

and

$$N_{xy} = N_{yx} = q\psi\left(\frac{\partial u}{\partial y} + \frac{\partial v}{\partial x}\right) \tag{10.28c}$$

where

$$\psi = \frac{Et}{(1 - \nu^2)} \quad \text{and} \quad q = \frac{1 - \nu}{2}$$

The stress-moments of Eq. (10.4) and the above stress resultants, when

* These are the tangential components of displacement of points lying on the 'mid-plane', as shown in Figs. 10.8 and 10.9.

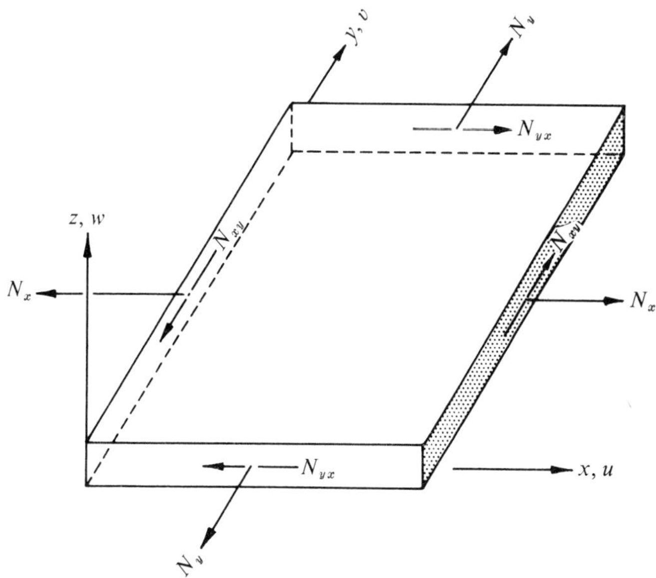

Fig. 10.9 'Stress resultants' in shell structures.

organized into a single matrix equation, lead to

$$\begin{Bmatrix} N_x \\ N_y \\ N_{xy} \\ M_x \\ M_y \\ M_{xy} \end{Bmatrix} = \begin{bmatrix} \psi & v\psi & 0 & 0 & 0 & 0 \\ v\psi & \psi & 0 & 0 & 0 & 0 \\ 0 & 0 & q\psi & 0 & 0 & 0 \\ 0 & 0 & 0 & D & vD & 0 \\ 0 & 0 & 0 & vD & D & 0 \\ 0 & 0 & 0 & 0 & 0 & qD \end{bmatrix} \begin{Bmatrix} \dfrac{\partial u}{\partial x} \\ \dfrac{\partial v}{\partial y} \\ \dfrac{\partial u}{\partial y} + \dfrac{\partial v}{\partial x} \\ -\dfrac{\partial^2 w}{\partial x^2} \\ -\dfrac{\partial^2 w}{\partial y^2} \\ -2\dfrac{\partial^2 w}{\partial x\,\partial y} \end{Bmatrix} \tag{10.29}$$

The vector on the left hand side of Eq. (10.29) lists the generalized 'stresses' within the plate while that on the right hand side the generalized 'strains' within it. Then, by definition, the square matrix in Eq. (10.29) is the 'elasticity' matrix [d] of the plate.

The primary matrices

For deriving the primary matrices of the stretched plate element, we must consider the in-plane forces and displacements in addition to the bending forces and displacements. In other words, a stretched plate element has 5 degrees of freedom as in Fig. 10.11.

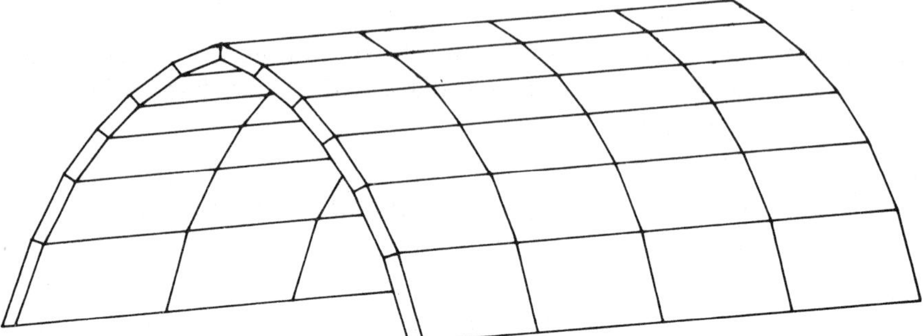

Fig. 10.10 A shell structure subdivided into a system of rectangular 'stretched plate elements'.

Consider, for example, the triangular stretched plate element: noting that the in-plane displacements $u(x, y)$ and $v(x, y)$ are mutually independent and that this element has 3 extremities, we can assume

$$u(x, y) = A_0 + A_1x + A_2y \tag{10.30a}$$

and

$$v(x, y) = A_3 + A_4x + A_5y \tag{10.30b}$$

in accordance with section 9.5(a) (the in-plane characteristics of elements will be discussed in detail in Chapter 11).

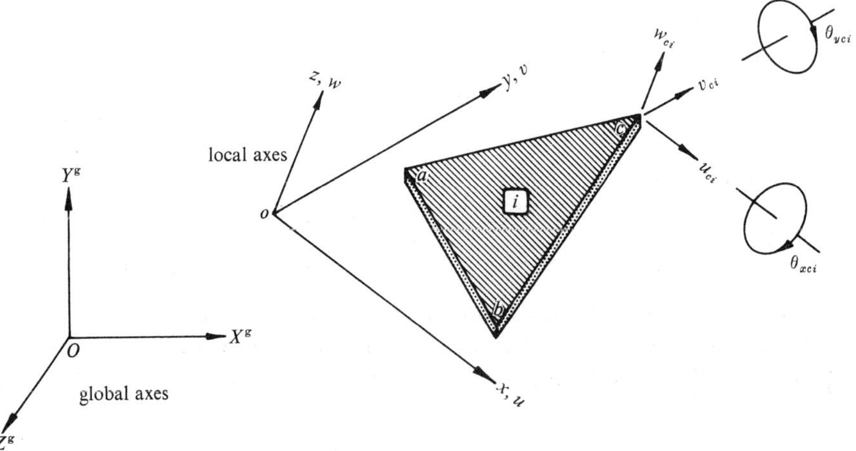

Fig. 10.11 A triangular 'stretched plate element'. (The displacements at one extremity of the element, assumed to lie on the x-y plane, are only shown).

Now, by virtue of the assumption of geometrical linearity, the transverse displacement $w(x, y)$ of the element is independent of both $u(x, y)$ and $v(x, y)$. For this reason $w(x, y)$ of Eq. (10.21) must now be redefined such that none of its constants is common with those of u or v. Let us therefore rewrite $w(x, y)$ merely by replacing the constants of Eq. (10.21) as follows:

$$w(x, y) = A_6 + A_7x + A_8y + \cdots + A_{13}(xy^2 + x^2y) + A_{14}y^3 \qquad (10.31)$$

Consequently, the constants of Eqs. (10.22) will also have to be altered accordingly. As usual, we shall write u, v and w in the matrix form of Eq. (10.8) where now

$$\{U\} = \begin{Bmatrix} u(x, y) \\ v(x, y) \\ w(x, y) \end{Bmatrix}, \qquad \{A\} = \{A_0 \quad A_1 \quad A_2 \quad \cdots \quad A_{13} \quad A_{14}\}^{\mathrm{T}}$$

and

$$[M] = \begin{bmatrix} 1 & x & y & 0 & 0 & 0 & 0 & 0 & 0 & 0 & 0 & 0 & 0 & 0 & 0 \\ 0 & 0 & 0 & 1 & x & y & 0 & 0 & 0 & 0 & 0 & 0 & 0 & 0 & 0 \\ 0 & 0 & 0 & 0 & 0 & 0 & 1 & x & y & x^2 & xy & y^2 & x^3 & (xy^2 + x^2y) & y^3 \end{bmatrix} \qquad (10.32)$$

The $[c]$ matrix of the element is now obtained by substituting its extremity coordinates into the expressions for u, v, w, θ_x and θ_y (with altered constants for θ_x and θ_y), in this order. Table 10.6 gives the $[c]$ matrix obtained in this way from Eqs. (10.22), (10.30) and (10.31).

Finally, the primary matrix $[N]$ is obtained by differentiating the expressions for u, v and w according to the scheme of the right hand vector of Eq. (10.29). It can be verified that

$$[N] = \begin{bmatrix} 0 & 1 & 0 & 0 & 0 & 0 & 0 & 0 & 0 & 0 & 0 & 0 & 0 & 0 & 0 \\ 0 & 0 & 0 & 0 & 0 & 1 & 0 & 0 & 0 & 0 & 0 & 0 & 0 & 0 & 0 \\ 0 & 0 & 1 & 0 & 1 & 0 & 0 & 0 & 0 & 0 & 0 & 0 & 0 & 0 & 0 \\ 0 & 0 & 0 & 0 & 0 & 0 & 0 & 0 & 0 & -2 & 0 & 0 & -6x & -2y & 0 \\ 0 & 0 & 0 & 0 & 0 & 0 & 0 & 0 & 0 & 0 & 0 & -2 & 0 & -2x & -6y \\ 0 & 0 & 0 & 0 & 0 & 0 & 0 & 0 & 0 & 0 & -2 & 0 & 0 & -4(x + y) & 0 \end{bmatrix} \qquad (10.33)$$

The primary matrices found in this way, together with $[d]$ (which is the square matrix of Eq. (10.29)) will now be substituted into Eq. (10.14) for calculating the $[K_i]$ of the triangular 'stretched plate element' i. A multiplication will show that in this case

$$[\bar{K}] = \begin{bmatrix} K' & 0 \\ 0 & K'' \end{bmatrix} \qquad (10.34)$$

Table 10.6 The $[c]$ matrix of a triangular 'stretched plate element'

$[c] =$

	1	2	3	4	5	6	7	8	9	10	11	12	13	14	15
1	1	x_a	y_a	0	0	0	0	0	0	0	0	0	0	0	0
2	0	0	0	1	x_a	y_a	0	0	0	0	0	0	0	0	0
3	0	0	0	0	0	0	1	x_a	y_a	x_a^2	$x_a y_a$	y_a^2	x_a^3	$x_a y_a^2 + x_a^2 y_a$	y_a^3
4	0	0	0	0	0	0	0	0	1	0	x_a	$2y_a$	0	$2x_a y_a + x_a^2$	$3y_a^2$
5	0	0	0	0	0	0	0	-1	0	$-2x_a$	$-y_a$	0	$-3x_a^2$	$-(y_a^2 + 2x_a y_a)$	0
6	1	x_b	y_b	0	0	0	0	0	0	0	0	0	0	0	0
7	0	0	0	1	x_b	y_b	0	0	0	0	0	0	0	0	0
8	0	0	0	0	0	0	1	x_b	y_b	x_b^2	$x_b y_b$	y_b^2	x_b^3	$x_b y_b^2 + x_b^2 y_b$	y_b^3
9	0	0	0	0	0	0	0	0	1	0	x_b	$2y_b$	0	$2x_b y_b + x_b^2$	$3y_b^2$
10	0	0	0	0	0	0	0	-1	0	$-2x_b$	$-y_b$	0	$-3x_b^2$	$-(y_b^2 + 2x_b y_b)$	0
11	1	x_c	y_c	0	0	0	0	0	0	0	0	0	0	0	0
12	0	0	0	1	x_c	y_c	0	0	0	0	0	0	0	0	0
13	0	0	0	0	0	0	1	x_c	y_c	x_c^2	$x_c y_c$	y_c^2	x_c^3	$x_c y_c^2 + x_c^2 y_c$	y_c^3
14	0	0	0	0	0	0	0	0	1	0	x_c	$2y_c$	0	$2x_c y_c + x_c^2$	$3y_c^2$
15	0	0	0	0	0	0	0	-1	0	$-2x_c$	$-y_c$	0	$-3x_c^2$	$-(y_c^2 + 2x_c y_c)$	0

where $[K'']$ is equal to the $[\bar{K}]$ matrix of Table 10.4 while

$$[K'] = \Delta\psi \begin{bmatrix} 0 & 0 & 0 & 0 & 0 & 0 \\ 0 & 1 & 0 & 0 & 0 & \nu \\ 0 & 0 & q & 0 & q & 0 \\ 0 & 0 & 0 & 0 & 0 & 0 \\ 0 & 0 & q & 0 & q & 0 \\ 0 & \nu & 0 & 0 & 0 & 1 \end{bmatrix} \tag{10.35}$$

The primary matrices of the rectangular 'stretched plate element' can also be derived in exactly the same way, by using the polynomial of Eq. (10.6) and with

$$u(x, y) = A_0 + A_1x + A_2y + A_3xy \tag{10.36a}$$

and

$$v(x, y) = A_4 + A_5x + A_6y + A_7xy \tag{10.36b}$$

The primary matrices together with the $[d]$ matrix defined by Eq. (10.29) will then be used, as usual, to calculate $[K_i]$.

Problem solution

Having calculated the $[K_i]$'s of all elements into which the plate is subdivided, they will be assembled, as usual, by the methods of Chapter 3 to form the overall stiffness matrix of the entire plate. The problem will then be solved in terms of Eq. (3.9) in which the listings of $\{\delta\}$ and $\{P\}$ are now as follows:

$$\{\delta\} = \{u_1 \ v_1 \ w_1 \ \theta_{x1} \ \theta_{y1} \ \cdots \ u_N \ v_N \ w_N \ \theta_{xN} \ \theta_{yN}\}^{\mathrm{T}}$$

and

$$\{P\} = \{F_1 \ Q_1 \ H_1 \ M_{x1} \ M_{y1} \ \cdots \ F_N \ Q_N \ H_N \ M_{xN} \ M_{yN}\}^{\mathrm{T}}$$

in which N is the total number of nodes in the discretized plate. For the kth node ($k = 1, 2, \ldots N$),

F_k, Q_k, H_k = externally applied forces at node k along x, y and z respectively.

u_k, v_k, w_k = displacements at node k along x, y and z respectively,

M_{xk}, M_{yk} = externally applied moments at node k,

θ_{xk}, θ_{yk} = rotations at node k.

10.9 Analysis of shells, folded roofs etc.

Both in-plane and transverse forces and displacements generally occur within a shell, a folded roof or a similar curved structure, when it is subjected to external forces. We may therefore analyze 'thin' structures of this type by subdividing them into 'stretched plate elements' discussed in the last section.

Fig. 10.10 shows an example of this, in which a shell has been subdivided into a system of rectangular stretched plate elements.

As usual, the $[K_i]$ matrix of the stretched plate element refers to its 'local' coordinates, e.g., *oxyz* in Fig. 10.11. However, as the elements into which such a structure is subdivided usually have different orientations, the $[K_i]$'s of individual elements must be transformed so that they all refer to a common 'global' system of coordinates. We have already encountered similar transformations in Chapters 6 and 7. For triangular and rectangular elements, the appropriate transformation matrices are

$$[T_i] = \begin{bmatrix} t_i & 0 & 0 \\ 0 & t_i & 0 \\ 0 & 0 & t_i \end{bmatrix} \quad \text{and} \quad [T_i] = \begin{bmatrix} t_i & 0 & 0 & 0 \\ 0 & t_i & 0 & 0 \\ 0 & 0 & t_i & 0 \\ 0 & 0 & 0 & t_i \end{bmatrix}$$

respectively, where

$$[t_i] \begin{bmatrix} x^g x & x^g y & x^g z & 0 & 0 \\ y^g x & y^g y & y^g z & 0 & 0 \\ z^g x & z^g y & z^g z & 0 & 0 \\ 0 & 0 & 0 & x^g x & x^g y \\ 0 & 0 & 0 & y^g x & y^g y \\ 0 & 0 & 0 & z^g x & z^g y \end{bmatrix}$$

Here $x^g y$, for instance, is the cosine of the angle between the X^g and y axes etc. (see Fig. 10.11). The 'global element stiffness matrix' $[K_i^g]$ of element i will then be given by

$$[K_i^g] = [T_i][K_i][T_i]^{\mathrm{T}} \tag{10.37}$$

in which $[K_i]$ refers to the 'stretched plate element' i. Also, note that in deriving Eq. (10.37) we have made use of the fact that the above $[T_i]$'s satisfy equations similar to Eq. (7.6).

The $[K_i^g]$'s will then be partitioned into their $S^g(i, J, K)$'s, whose assembly by the methods of Chapter 3 will result in the overall stiffness matrix $[K^g]$. The problem can then be solved in terms of Eq. (6.14), having accounted for all the prescribed boundary conditions.

Since $[t_i]$ here is rectangular, it is easily verified from Eq. (10.37) that the 'stretched plate element' has 6 degrees of freedom in the global axes although it has 5 degrees of freedom in its local axes (recall that a similar situation arose in Chapter 7 where a single local degree of freedom resulted in more than one global degrees of freedom). It may also be verified that the extra global degree of freedom is in fact the moment M_z^g which lies on a plane normal to the Z^g

axis. Consequently, if the discretized structure contains a total of N nodes, then

$$\{P^g\} = \{F_1^g\ Q_1^g\ H_1^g\ M_{x1}^g\ M_{y1}^g\ M_{z1}^g \cdots F_N^g\ Q_N^g\ H_N^g\ M_{xN}^g\ M_{yN}^g\ M_{zN}^g\}^{\mathrm{T}}$$

and

$$\{\delta^g\} = \{u_1^g\ v_1^g\ w_1^g\ \theta_{x1}^g\ \theta_{y1}^g\ \theta_{z1}^g \cdots u_N^g\ v_N^g\ w_N^g\ \theta_{xN}^g\ \theta_{yN}^g\ \theta_{zN}^g\}^{\mathrm{T}}$$

All quantities in these equations are the global counterparts of those of $\{P\}$ and $\{\delta\}$ in section 10.8, except M_{zk}^g which denotes the external global moment about Z^g applied at node k, and θ_{zk}^g denotes the corresponding rotation of k.

10.10 Plate-beam systems

The analysis of plate-beam, shell-ring beam and similar combinations does not present any difficulty. Consider, for instance, the plate-beam system of Fig. 10.12a, a finite element subdivision of which is shown in Fig. 10.12b. The $[K]$ matrix for this combination is assembled as usual; applying the method of section 3.4 to node 2, for instance, we can show that the contribution of this node to Eq. (3.9) for this structure is

$$\begin{array}{c|cccccc} & 1 & 2 & 3 & 4 & 5 & 6 \\ \hline & S^p(1,3,2) & S^p(1,3,3) & S^p(2,2,3) & S^p(1,3,1) & S^p(1,3,4) & S^p(2,2,4) \\ 2 & +\,S^b(1,2,1) & +\,S^p(2,2,2) & +\,S^b(2,1,2) & & +\,S^p(2,2,1) & \\ & & +\,S^b(1,2,2) & & & & \\ & & +\,S^b(2,1,1) & & & & \end{array} \begin{Bmatrix} \vdots \\ w_2 \\ \theta_{x2} \\ \theta_{y2} \\ \vdots \end{Bmatrix} = \begin{Bmatrix} \vdots \\ H_2 \\ M_{x2} \\ M_{y2} \\ \vdots \end{Bmatrix}$$

where the directions of forces and displacements are as shown in Fig. 10.12c. Here, $S^p(i, J, K)$ and $S^b(i, J, K)$ are respectively the $S(i, J, K)$'s of the rectangular plate element (section 10.3) and the beam element of Chapter 2 (note that in this case M_{y2} acts on the beam as a torsional load). The computer is easily programmed to assemble the complete $[K]$ and $\{P\}$ matrices in this way automatically.

10.11 Observations

In section 10.3 we saw that while the transverse deflection of the element was continuous across all interfaces between adjacent elements, its rotation about an axis was, however, discontinuous across interfaces parallel to that axis. Consequently, the assumed transverse deflection of Eq. (10.6) is said to be 'non-conforming'. By using Hermitian interpolation formulae, Bogner *et al.*[3] have derived 'conforming' deflection functions which do not lead to such discontinuities. In general, accuracy due to these functions is, however, inferior to that achieved by using the non-conforming deflections.

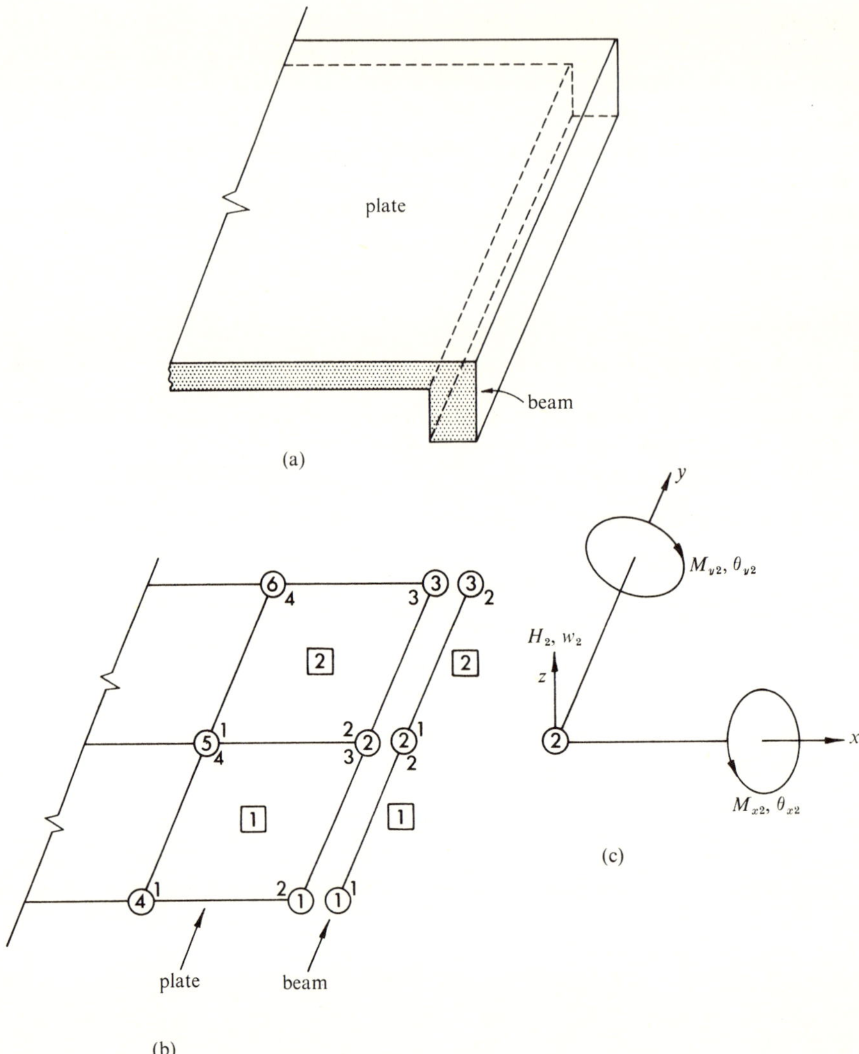

Fig. 10.12 (a) A plate-beam combination. (b) A finite element subdivision of (a). (c) Externally applied nodal forces.

The elements of this chapter cannot be used to analyze 'thick' plates and shells, defined as those in which t is not small compared with the other dimensions, since the stresses and displacements within these are essentially three dimensional. The elements of Chapter 12 would have to be used to analyze them.

To the beginner, the size of the primary matrices of this chapter, particularly those for dealing with stretched plates, shells etc. (sections 10.8 and 10.9), is likely

to appear formidable. We must emphasize, however, that any dismay about this is completely unfounded, since in principle, the computer processing of these matrices is no different from those of much simpler elements and does not present any significant additional difficulty. A simple program to deal with flat plate problems without the in-plane aspect will be given in section 15.5. As an exercise, the reader is advised to adapt this program to shell and similar structures by introducing into it the necessary operations of the transformation matrices given in section 10.9.

References

1. Timoshenko, S. and Woinowsky-Krieger, S., *Theory of plates and Shells*, 2nd edn. McGraw-Hill, 1959.

2. Zienkiewicz, O. C. and Cheung, Y. K., 'The finite element method for analysis of elastic isotropic and orthotropic slabs', *Proc. Inst. Civ. Eng.* **28,** 471–88, 1964.

3. Bogner, F. K., Fox, R. L., Schmit, L. A., 'The generation of inter element-compatible stiffness and mass matrices by the use of interpolation formulae', *Proc. Conf. Matrix Methods in Struct. Mech.*, Air Force Base Institute of Technology, Wright Patterson A.F. Base, Ohio, 1965.

11 Analysis of plane problems

The finite element method provides a simple and accurate means for the solution of plane problems (sections 8.5 and 8.6) of Elasticity, by using two-dimensional elements of various shapes. Elements of this type were first successfully used by Clough[1] and Turner *et al.*[2] These elements are generally referred to as 'in-plane' elements, since the displacements and forces in them occur on the plane of the element itself. We have already come across such elements in section 10.8.

11.1 The triangular element

Figure 11.1 shows the extremity forces and displacements associated with a typical triangular plane element. As the element has three extremities, the displacement components, u and v, within it can be assumed to vary as in Eq. (10.30), in which each polynomial is defined by three constants.* Introducing

$$\{U\} = \begin{Bmatrix} u(x, y) \\ v(x, y) \end{Bmatrix},$$

we can write Eqs. (10.30) in the matrix form as

$$\{U\} = [M]\{A\} \tag{11.1a}$$

where

$$\{A\} = \{A_0 \quad A_1 \quad \cdots \quad A_5\}^{\mathrm{T}}$$

and

$$[M] = \begin{bmatrix} 1 & x & y & 0 & 0 & 0 \\ 0 & 0 & 0 & 1 & x & y \end{bmatrix} \tag{11.1b}$$

Following the method of section 9.8(c), we can demonstrate that A_0 and A_3 represent the element's rigid-body displacements along the x and y axes respectively.

The displacements at each extremity of the element are now obtained by substituting the coordinates of that extremity into Eq. (10.30). Thus, at extremity a, for example,

$$u_{ai} = u(x_a, y_a) = A_0 + A_1 x_a + A_2 y_a$$
$$v_{ai} = v(x_a, y_a) = A_3 + A_4 x_a + A_5 y_a$$

* See section 9.5(a).

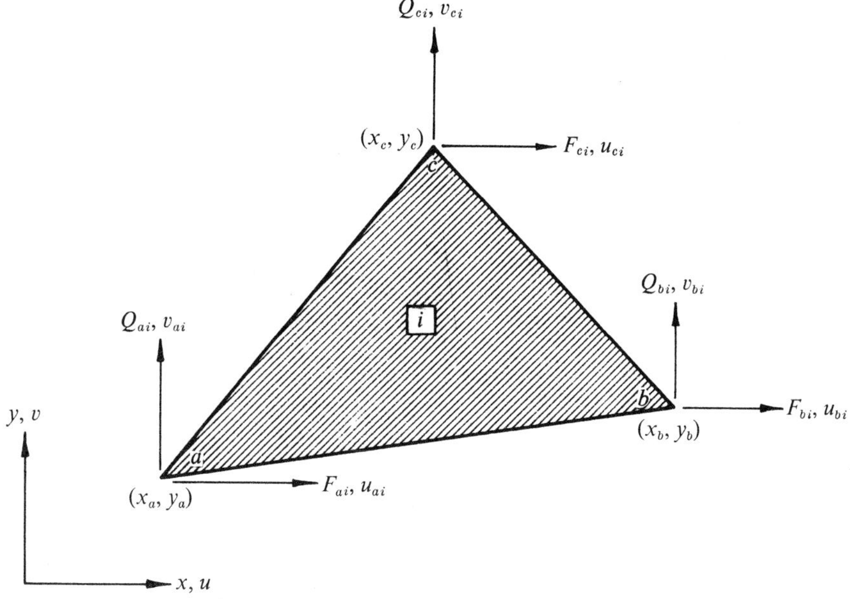

Fig. 11.1 The general triangular in-plane element; uniform thickness $= t$.

etc. We shall now put all these extremity displacements in the following matrix form:

$$\{\delta\} = [c]\{A\} \tag{11.2}$$

where
$$\{\delta\} = \{u_{ai} \quad v_{ai} \quad u_{bi} \quad v_{bi} \quad u_{ci} \quad v_{ci}\}^{\mathsf{T}}$$

and
$$[c] = \begin{bmatrix} 1 & x_a & y_a & 0 & 0 & 0 \\ 0 & 0 & 0 & 1 & x_a & y_a \\ 1 & x_b & y_b & 0 & 0 & 0 \\ 0 & 0 & 0 & 1 & x_b & y_b \\ 1 & x_c & y_c & 0 & 0 & 0 \\ 0 & 0 & 0 & 1 & x_c & y_c \end{bmatrix} \tag{11.3}$$

From Eqs. (8.1), (8.3) and (8.4a), the strains $\{\epsilon\}$ within the element are

$$\{\epsilon\} = \begin{Bmatrix} e_{xx} \\ e_{yy} \\ e_{xy} \end{Bmatrix} = \begin{Bmatrix} \dfrac{\partial u}{\partial x} \\ \dfrac{\partial v}{\partial y} \\ \dfrac{\partial u}{\partial y} + \dfrac{\partial v}{\partial x} \end{Bmatrix} \tag{11.4a}$$

The primary matrix $[N]$ can now be obtained by differentiating the displacements of Eqs. (10.30) according to the scheme of Eq. (11.4a); we can demonstrate that in fact

$$\{\epsilon\} = \begin{Bmatrix} A_1 \\ A_5 \\ A_2 + A_4 \end{Bmatrix} = \begin{bmatrix} 0 & 1 & 0 & 0 & 0 & 0 \\ 0 & 0 & 0 & 0 & 0 & 1 \\ 0 & 0 & 1 & 0 & 1 & 0 \end{bmatrix} \{A\}$$

or symbolically,

$$\{\epsilon\} = [N]\{A\} \tag{11.4b}$$

where $[N]$ denotes the above rectangular matrix. Having thus determined the explicit forms of the three primary matrices $[M]$, $[N]$ and $[c]$, we can derive the various characteristics of the element from the formulae of section 9.8; it is easily shown, for instance, that the element stiffness matrix is given by

$$[K_i] = t\,.\,[c^{-1}]^{\mathrm{T}}[\bar{K}][c^{-1}] \tag{11.5}$$

where

$$[\bar{K}] = \int_\Delta [N]^{\mathrm{T}}[d][N]\,\mathrm{d}x\,\mathrm{d}y \tag{11.6}$$

The above integration will be carried out over the entire area of the element, whose uniform thickness is t. Equation (11.4b) shows that $[N]$ is not a function of the current coordinates; as a result, for the triangular element we obtain*

$$[\bar{K}] = [N]^{\mathrm{T}}[d][N]\int_\Delta \mathrm{d}x\,\mathrm{d}y = \Delta[N]^{\mathrm{T}}[d][N] \tag{11.7}$$

where Δ = area of element i and $\int_\Delta$ denotes integration over a triangular area. For plane stress the $[d]$ matrix to be used in Eq. (11.7) is that of Eq. (8.19) while, for plane strain, we must take $t = 1$ and use the $[d]$ matrix of Eq. (8.18).

Since the computer will be used for calculations, it is superfluous to write down the explicit forms of the $S(i, J, K)$'s of $[K_i]$; instead, it is more convenient to instruct the computer to form $[K_i]$ from Eqs. (11.5) and (11.7), and then identify the $S(i, J, K)$'s from it automatically. There is one minor difficulty however; as $[c]$ here has a number of zeros on its leading diagonal, it cannot be inverted directly. This is easily overcome by inverting $[c]$ indirectly, by the methods given in section 15.3.

Alternatively, a solution of the simultaneous equations represented by Eq. (11.2) gives

$$A_0 = \frac{1}{2\Delta}((x_b y_c - x_c y_b)u_{ai} + (x_c y_a - x_a y_c)u_{bi} + (x_a y_b - x_b y_a)u_{ci}),$$

$$A_1 = \frac{1}{2\Delta}(y_{bc}u_{ai} + y_{ca}u_{bi} + y_{ab}u_{ci})$$

and

$$A_2 = \frac{1}{2\Delta}(x_{cb}u_{ai} + x_{ac}u_{bi} + x_{ba}u_{ci})$$

* See Appendix 3 for integration over a triangular area.

in which $2\Delta = (y_{bc}x_{ab} - y_{ab}x_{bc})$ = twice the area of triangle i. Note that the symbols x_{ab}, y_{bc} etc. denote 'coordinate differences', defined as

$$x_{ab} = x_a - x_b$$

$$y_{bc} = y_b - y_c$$

etc. Also, it is obvious that $x_{ab} = -x_{ba}$, $y_{bc} = -y_{cb}$ etc.

The constants A_3, A_4 and A_5 will be given by the expressions for A_0, A_1 and A_2 respectively, when u_{ai}, u_{bi} and u_{ci} are replaced by v_{ai}, v_{bi} and v_{ci} respectively in these expressions. Arranging these constants in the matrix form of

$$\{A\} = [X]\{\delta\},$$

we observe that by definition (section 9.8). $[X] = [c^{-1}]$. Table 11.1 gives $[c^{-1}]$ obtained in this way, which really amounts to an indirect inversion of $[c]$.

The finite element analysis of plane problems by triangular elements is initiated by forming the $[K_i]$'s of individual elements from Eq. (11.5); next, the $S(i, J, K)$'s of elements are assembled, in the manner indicated in Chapter 3,

Table 11.1 The $[c]^{-1}$ matrix for the triangular element

$$[c]^{-1} = \frac{1}{2\Delta}\begin{bmatrix} x_by_c - x_cy_b & 0 & x_cy_a - x_ay_c & 0 & x_ay_b - x_by_a & 0 \\ y_{bc} & 0 & y_{ca} & 0 & y_{ab} & 0 \\ x_{cb} & 0 & x_{ac} & 0 & x_{ba} & 0 \\ 0 & x_by_c - x_cy_b & 0 & x_cy_a - x_ay_c & 0 & x_ay_b - x_by_a \\ 0 & y_{bc} & 0 & y_{ca} & 0 & y_{ab} \\ 0 & x_{cb} & 0 & x_{ac} & 0 & x_{ba} \end{bmatrix}$$

to form the overall stiffness matrix $[K]$ for the entire discretized body. Finally, the prescribed boundary conditions are introduced into $[K]$, upon which the problem can be solved in terms of Eq. (3.9). The listings of $\{P\}$ and $\{\delta\}$ here are as follows:

$$\{P\} = \{F_1 \quad Q_1 \quad F_2 \quad Q_2 \quad \cdots \quad F_N \quad Q_N\}^{\mathrm{T}} \tag{11.8}$$

and

$$\{\delta\} = \{u_1 \quad v_1 \quad u_2 \quad v_2 \quad \cdots \quad u_N \quad v_N\}^{\mathrm{T}} \tag{11.9}$$

where N is the total number of nodes in the discretized body and, at node k $(= 1, 2, \ldots N)$

F_k = external force along x at k

Q_k = external force along y at k and

u_k, v_k = displacements along x and y of node k, respectively

Thermal, initial strain and body forces, when present, must be included in $\{P\}$ in the manner indicated in section 9.8. A simple computer program for

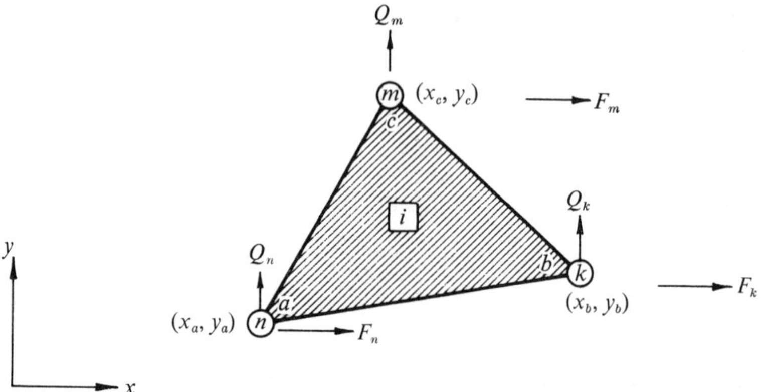

Fig. 11.2 A triangular in-plane element.

solving plane problems by triangular elements will be given in section 15.5.

The stresses within individual elements are found from computed nodal displacements by eliminating $\{\epsilon\}$ between the constitutive equation of the element and Eq. (11.4b). Thus, in the case of plane stress, for example, the stresses caused by elastic, thermal and initial strains within a typical element (Fig. 11.2) will be given by

$$\begin{Bmatrix} \sigma_{xx} \\ \sigma_{yy} \\ \tau_{xy} \end{Bmatrix} = [d][N][c^{-1}]\begin{Bmatrix} u_n \\ v_n \\ u_k \\ v_k \\ u_m \\ v_m \end{Bmatrix} - \frac{E\alpha T}{(1-\nu)}\begin{Bmatrix} 1 \\ 1 \\ 0 \end{Bmatrix} - [d]\begin{Bmatrix} e_{\mathrm{i}xx} \\ e_{\mathrm{i}yy} \\ e_{\mathrm{i}xy} \end{Bmatrix} \tag{11.10}$$

in which $[d]$ refers to Eq. (8.19) and the nodes n, k and m are assumed to be attached to the extremities a, b and c respectively. The principal stresses, corresponding to the above stresses, can now be determined from the methods of section 8.10, if required.

Since both $[c]$ and $[N]$ are independent of the current coordinates, Eq. (11.10) shows that each component of stress (and hence strain) has a constant value throughout the element. Consequently, the stress equilibrium equations (Eqs. 8.21) are satisfied identically.

Thermal forces

For the typical element of Fig. 11.2, the thermal force vector $\{P_{\mathrm{t}}\}$ will be obtained by substituting for $\{\sigma\}_{\mathrm{t}}$ from the constitutive equations into Eq. (9.12).

Thus, for plane strain, for example (Eq. 8.18),

$$\{P_t\} = \begin{Bmatrix} F_n \\ Q_n \\ F_k \\ Q_k \\ F_m \\ Q_m \end{Bmatrix} = \frac{E\alpha T}{(1-2\nu)}[c^{-1}]^{\mathrm{T}}[N]^{\mathrm{T}}\begin{Bmatrix} 1 \\ 1 \\ 0 \end{Bmatrix}\int_{\Delta} \mathrm{d}x\,\mathrm{d}y = \frac{E\alpha T}{(2-4\nu)}\begin{Bmatrix} y_{bc} \\ x_{cb} \\ y_{ca} \\ x_{ac} \\ y_{ab} \\ x_{ba} \end{Bmatrix}$$

in which the thickness t has been taken as unity and we have used $[c^{-1}]$ of Table 11.1. The components of $\{P_t\}$ must be applied as external nodal forces at n, k and m, in addition to other external forces that these nodes may have (see section 9.8).

Body forces

Equation (9.19) shows that the body force vector is not a function of the elasticity matrix. Therefore, for a given element shape, $\{P_b\}$ will be the same for both plane stress and plane strain, except for t, which will be equal to unity for the latter. Now, for a plane element lying on the $x - y$ plane, we have from section 9.8

$$\{P_b\} = \rho t \,.\, [c^{-1}]^{\mathrm{T}} \int_{\Delta} [M]^{\mathrm{T}} \begin{Bmatrix} \bar{X} \\ \bar{Y} \end{Bmatrix} \mathrm{d}x\,\mathrm{d}y \qquad (11.11)$$

in which the integration is performed over the entire area of the element.

As an example, let us now derive the body force vector for the element of Fig. 11.2. Assuming that the body forces here are due only to gravity, so that $\bar{X} = 0$ and $\bar{Y} = -g$ where g is the gravity acceleration, we can show from Eqs. (11.1b) and (11.11) that

$$\{P_b\} = \rho t \,.\, [c^{-1}]^{\mathrm{T}} \int_{\Delta} \begin{bmatrix} 1 & 0 \\ x & 0 \\ y & 0 \\ 0 & 1 \\ 0 & x \\ 0 & y \end{bmatrix} \begin{Bmatrix} 0 \\ -g \end{Bmatrix} \mathrm{d}x\,\mathrm{d}y = -\rho t g [c^{-1}]^{\mathrm{T}} \int_{\Delta} \begin{Bmatrix} 0 \\ 0 \\ 0 \\ 1 \\ x \\ y \end{Bmatrix} \mathrm{d}x\,\mathrm{d}y \qquad (11.12)$$

Now, using the integrals of Appendix 3 (Eqs. A3.8a, b and c) and $[c^{-1}]$ given

in Table 11.1, we can show that Eq. (11.12) leads to

$$\{P_b\} = \begin{Bmatrix} F_n \\ Q_n \\ F_k \\ Q_k \\ F_m \\ Q_m \end{Bmatrix} = -\rho g t\, \Delta[c^{-1}]^T \begin{Bmatrix} 0 \\ 0 \\ 0 \\ 1 \\ 0 \\ 0 \end{Bmatrix} = -\frac{\rho g t}{2} \begin{Bmatrix} 0 \\ x_b y_c - x_c y_b \\ 0 \\ x_c y_a - x_a y_c \\ 0 \\ x_a y_b - x_b y_a \end{Bmatrix} \quad (11.13)$$

As the integrals of Eqs. (A3.8) are with respect to the centroid of the triangle as origin, the extremity coordinates above must be measured from the centroid too (alternatively, Eqs. (A3.7) may be used).

It is interesting to verify by a simple calculation that for any triangle, the non-zero components of $\{P_b\}$, given by Eq. (11.13) are

$$Q_n = Q_k = Q_m = -\frac{\rho g t \Delta}{3}$$

thus indicating that one-third of the total weight of the triangle is sustained by each node attached to it. This is what we might have anticipated from a guess based on 'rule of thumb'. Body forces due to inertia, on the other hand, give rise to 'mass' matrices (section 14.1) whose physical interpretation is more involved.

Interface continuity

By substituting the expressions for A_0 A_1 and A_2 into Eqs. (10.30) and noting that in Fig. 11.2

$$u_{ai} = u_n, \qquad u_{bi} = u_k \quad \text{and} \quad u_{ci} = u_m$$

we can demonstrate that

$$u(x, y) = a_1 u_n + a_2 u_k + a_3 u_m \quad (11.14a)$$

where

$$a_1 = \frac{1}{2\Delta}((y_{bc}(x - x_c) - x_{bc}(y - y_c))$$

$$a_2 = \frac{1}{2\Delta}((y_{ca}(x - x_a) - x_{ca}(y - y_a))$$

and

$$a_3 = \frac{1}{2\Delta}((y_{ab}(x - x_b) - x_{ab}(y - y_b))$$

Similarly,

$$v(x, y) = a_1 v_n + a_2 v_k + a_3 v_m \quad (11.14b)$$

We can now show from Eqs. (11.14) that the variations of $u(x, y)$ and $v(x, y)$ along any edge of the triangle depends only on the displacements of the nodes defining that particular edge. This can be done as follows: the equation of the line defining the edge $k - m$ (Fig. 11.2), for example, is

$$y = \frac{1}{x_{bc}}(y_{bc}x + x_b y_c - x_c y_b)$$

Substitution of y into Eqs. (11.14) shows that the displacements along $k - m$ vary as

$$u(x, y) = a_2 u_k + a_3 u_m$$

and

$$v(x, y) = a_2 v_k + a_3 v_m$$

This implies that when any edge defines a common boundary between two adjacent triangular elements, the displacements across this boundary are continuous.

A differentiation of Eqs. (10.30) shows that the slopes of $u(x, y)$ and $v(x, y)$ are constant throughout the triangle; however, as each deformed element will normally have a different level of strain (slope), there will be a discontinuity of slopes across interfaces between adjacent elements.

11.2 The rectangular element

The extremity forces and displacements in the general rectangular in-plane element are as shown in Fig. 11.3. The study of the properties of this element is facilitated by introducing the non-dimensional coordinates, x' and y', defined as

$$x' = \frac{x}{h_x} \quad \text{and} \quad y' = \frac{y}{h_y}$$

Since the element has four extremities, the variation of each displacement component within it can be assumed as a polynomial defined by four constants (section 9.5(a)); in particular, let

$$u(x', y') = A_0 + A_1 x' + A_2 y' + A_3 x'y' \tag{11.15a}$$

and

$$v(x', y') = A_4 + A_5 x' + A_6 y' + A_7 x'y' \tag{11.15b}$$

Writing these equations in the matrix form of Eq. (11.1a), we have

$$\{U\} = \begin{Bmatrix} u(x', y') \\ v(x', y') \end{Bmatrix}$$

$$\{A\} = \{A_0 \quad A_1 \quad \cdots \quad A_7\}^{\mathrm{T}}$$

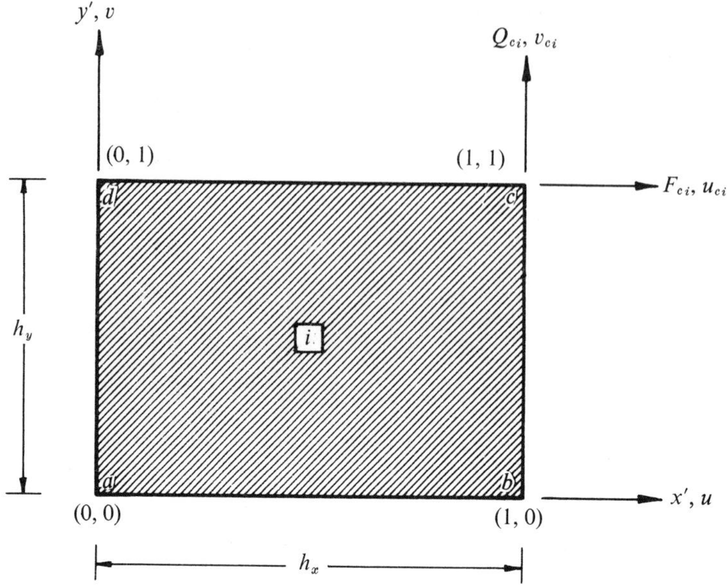

Fig. 11.3 The rectangular in-plane element; uniform thickness $= t$ (for clarity the extremity forces and displacements at extremity c only are shown).

and
$$[M] = \begin{bmatrix} 1 & x' & y' & x'y' & 0 & 0 & 0 & 0 \\ 0 & 0 & 0 & 0 & 1 & x' & y' & x'y' \end{bmatrix} \tag{11.16}$$

Notice that the term, $x'y'$, has been chosen in preference to others, as it ensures a linear variation of displacements along the edges of the element; clearly, each displacement is represented by a two-degree surface, and varies linearly along the y' and x' axes when $x' =$ constant and $y' =$ constant respectively. Consequently, the continuity of displacements across interfaces between adjacent rectangular elements is ensured, as we shall see later. Noting that the extremities a, b, c and d are located at the non-dimensional coordinates (0, 0), (1, 0), (1, 1) and (0, 1) respectively, we have from Eqs. (11.15)

$$u_{ai} = u(0, 0) = A_0$$
$$v_{ai} = v(0, 0) = A_4$$
$$u_{bi} = u(1, 0) = A_0 + A_1$$

etc. We shall now express the eight extremity displacements in the matrix form of Eq. (11.2); then, it may be verified that with

$$\{\delta\} = \{u_{ai} \quad v_{ai} \quad u_{bi} \quad v_{bi} \quad u_{ci} \quad v_{ci} \quad u_{di} \quad v_{di}\}^{\mathrm{T}}$$

the primary matrix $[c]$ is given by

$$[c] = \begin{bmatrix} 1 & 0 & 0 & 0 & 0 & 0 & 0 & 0 \\ 0 & 0 & 0 & 0 & 1 & 0 & 0 & 0 \\ 1 & 1 & 0 & 0 & 0 & 0 & 0 & 0 \\ 0 & 0 & 0 & 0 & 1 & 1 & 0 & 0 \\ 1 & 1 & 1 & 1 & 0 & 0 & 0 & 0 \\ 0 & 0 & 0 & 0 & 1 & 1 & 1 & 1 \\ 1 & 0 & 1 & 0 & 0 & 0 & 0 & 0 \\ 0 & 0 & 0 & 0 & 1 & 0 & 1 & 0 \end{bmatrix} \tag{11.17}$$

As in the triangular element, the strains within the element are

$$\{\epsilon\} = \begin{Bmatrix} e_{xx} \\ e_{yy} \\ e_{xy} \end{Bmatrix} = \begin{Bmatrix} \dfrac{1}{h_x} \cdot \dfrac{\partial u}{\partial x'} \\ \dfrac{1}{h_y} \cdot \dfrac{\partial v}{\partial y'} \\ \dfrac{1}{h_y} \cdot \dfrac{\partial u}{\partial y'} + \dfrac{1}{h_x} \cdot \dfrac{\partial v}{\partial x'} \end{Bmatrix} \tag{11.18}$$

The third primary matrix $[N]$ is obtained by differentiating Eqs. (11.15) according to the above scheme; expressing the result of this differentiation in the matrix form of Eq. (11.4b), we can show that

$$[N] = \begin{bmatrix} 0 & \dfrac{1}{h_x} & 0 & \dfrac{y'}{h_x} & 0 & 0 & 0 & 0 \\ 0 & 0 & 0 & 0 & 0 & 0 & \dfrac{1}{h_y} & \dfrac{x'}{h_y} \\ 0 & 0 & \dfrac{1}{h_y} & \dfrac{x'}{h_y} & 0 & \dfrac{1}{h_x} & 0 & \dfrac{y'}{h_x} \end{bmatrix} \tag{11.19}$$

Having thus determined the explicit forms of the three primary matrices for the assumed displacements of Eqs. (11.15), we can now derive the various element characteristics from the formulae of section 9.8. The element stiffness matrix, for example, can be found from Eq. (11.5) by introducing into it

$$[\bar{K}] = \int_0^1 \int_0^1 [N]^{\mathrm{T}} [d][N] \, \mathrm{d}x' \, \mathrm{d}y'$$

Here again, we are unable to invert $[c]$ directly, since it has zeros on its leading diagonal. An indirect inversion by the methods of section 15.3, on the other hand, gives

$$[c^{-1}] = \begin{bmatrix} 1 & 0 & 0 & 0 & 0 & 0 & 0 & 0 \\ -1 & 0 & 1 & 0 & 0 & 0 & 0 & 0 \\ -1 & 0 & 0 & 0 & 0 & 0 & 1 & 0 \\ 1 & 0 & -1 & 0 & 1 & 0 & -1 & 0 \\ 0 & 1 & 0 & 0 & 0 & 0 & 0 & 0 \\ 0 & -1 & 0 & 1 & 0 & 0 & 0 & 0 \\ 0 & -1 & 0 & 0 & 0 & 0 & 0 & 1 \\ 0 & 1 & 0 & -1 & 0 & 1 & 0 & -1 \end{bmatrix}$$

The remaining steps leading up to the solution of plane problems by rectangular elements, are identical to those for the triangular element. Equations (11.4b) and (11.19) show that the strains and hence the stresses within the element vary linearly; consequently, the stress equilibrium equations (Eqs. 8.21) are not satisfied.

Interface continuity

By eliminating the constants from Eqs. (11.15) and noting that in Fig. 11.4 the extremities a, b, c and d are attached to the nodes n, k, m and j respectively, we can demonstrate that the explicit forms of Eqs. (11.15) are

$$u(x', y') = (1 - x')(1 - y')u_n + (x' - x'y')u_k + x'y'u_m + (y' - x'y')u_j \quad (11.20a)$$

and

$$v(x', y') = (1 - x')(1 - y')v_n + (x' - x'y')v_k + x'y'v_m + (y' - x'y')v_j \quad (11.20b)$$

With $x' = 1$, for example, these displacements will vary along the edge $k - m$ as

$$u(1, y') = (1 - y')u_k + y'u_m$$

and

$$v(1, y') = (1 - y')v_k + y'v_m$$

As these displacements depend only on the nodal displacements of k and m, we conclude that the displacements of Eq. (11.15) are continuous across $k - m$, when $k - m$ forms a common interface between two adjacent rectangular elements. We can show in this way that the displacements of Eq. (11.15) are continuous across all interfaces between adjacent rectangular elements.

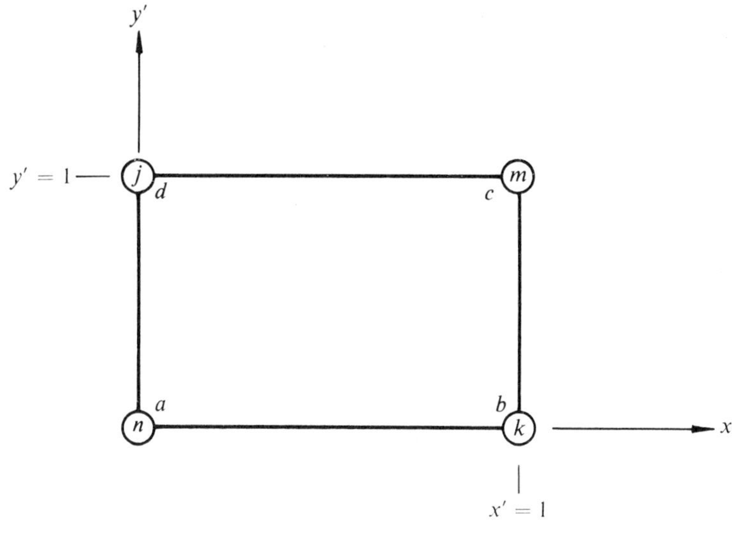

Fig. 11.4

Differentiation of Eqs. (11.20) with respect to x and y leads to

$$h_x e_{xx} = (y' - 1)u_n + (1 - y')u_k + y'u_m - y'u_j$$
$$h_y e_{yy} = (x' - 1)v_n + (1 - x')v_k + x'v_m - x'v_j$$

It is easily verified from these equations and Fig. 11.4 that

(a) e_{xx} and e_{yy} are constant along the lines $y' =$ constant and $x' =$ constant respectively, and

(b) e_{xx} is continuous across interfaces (between adjacent rectangular elements) parallel to the x' axis, since it depends only on the displacements of nodes defining those interfaces. For the same reason, e_{yy} is continuous across interfaces parallel to the y' axis.

It may be verified in this way that e_{xy}, on the other hand, is not continuous across any interface.

11.3 The quadrilateral element

The characteristics of the quadrilateral element (Fig. 11.5) can also be derived from assumed displacements similar to those of Eqs. (11.15). However, the displacements of Eqs. (11.15) will not, in general, guarantee the continuity of displacements across the interfaces between adjacent elements, owing to the inclination of these interfaces to the coordinate axes. This may be remedied by expressing the assumed displacements in terms of the coordinates of a non-orthogonal and non-dimensional system shown in Fig. 11.5. Quantities in the

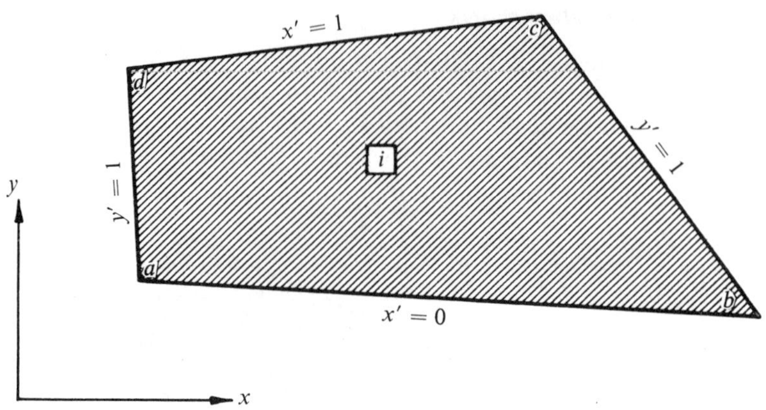

Fig. 11.5 The quadrilateral in-plane element; uniform thickness = t.

original orthogonal system *yox* are then transformed into this system by using the 'Jacobian of transformation'. By definition, the Jacobian determinant $|J|$ in this case is

$$|J| = \begin{vmatrix} \dfrac{\partial x}{\partial x'} & \dfrac{\partial y}{\partial x'} \\ \dfrac{\partial x}{\partial y'} & \dfrac{\partial y}{\partial y'} \end{vmatrix} = \frac{\partial x}{\partial x'} \cdot \frac{\partial y}{\partial y'} - \frac{\partial x}{\partial y'} \cdot \frac{\partial y}{\partial x'}$$

We can now determine the element stiffness matrix of the quadrilateral element from Eq. (11.5), by transforming Eq. (11.6) as

$$[\bar{K}] = \int_0^1 \int_0^1 [N]^{\mathrm{T}}[d][N]\,|J|\,\mathrm{d}x'\,\mathrm{d}y'$$

in which it is understood that $[N]$ has also been transformed by the Jacobian. As these integrands have polynomials as denominator, their closed form integration over the area of the quadrilateral is generally extremely difficult and usually impossible. Alternatively, numerical integration procedures can be employed to evaluate them; unfortunately, these entail lengthy computer programs[3]. This is the most unattractive feature of this element.

In view of these difficulties, it is a matter of some debate as to whether we could do just as well, if not better, by treating the quadrilateral as a combination of two triangular elements. This would seem to be a preferable alternative, at least until such time that further research has simplified the derivation of the characteristics of this element and has established justifiable evidence of better accuracy by its use. The problems posed by this element are nevertheless interesting and potentially useful.

11.4 The transformation matrix

It was pointed out in section 9.5 that when the 'local' axes of individual elements representing a given body are not parallel to each other, the 'local' forces, displacements and element characteristics must be transformed to the global axes. Such a situation is typified by the element of Fig. 11.6 in which the degrees

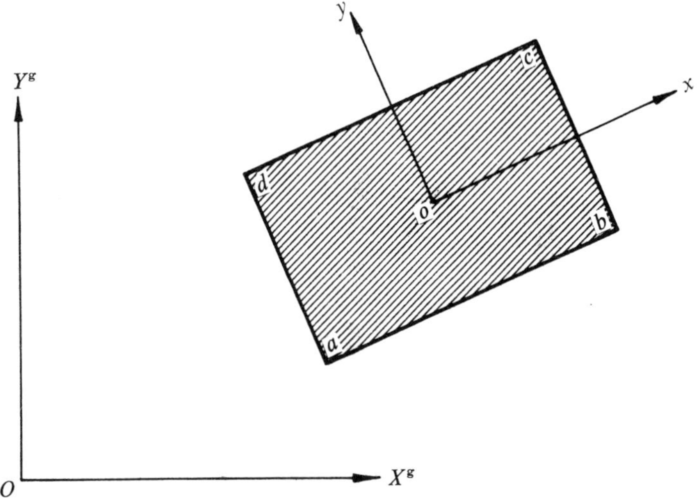

Fig. 11.6 The 'local' (yox) and 'global' (Y^gOX^g) axes of typical plane element.

of freedom are the same as those of Fig. 11.3. Following the method of Chapter 6, the reader may verify that the relevant transformation matrix here is

$$[T_i] = \begin{bmatrix} t_i & 0 & 0 & 0 \\ 0 & t_i & 0 & 0 \\ 0 & 0 & t_i & 0 \\ 0 & 0 & 0 & t_i \end{bmatrix} \tag{11.21}$$

where

$$[t_i] = \begin{bmatrix} x^g x & x^g y \\ y^g x & y^g y \end{bmatrix}$$

Here $x^g x$ denotes the cosine of the angle between X^g and x axes etc.

11.5 An example

Consider the following problem:

A rigid footing, shown in Fig. 11.7a, rests on an isotropic and homogeneous elastic medium which extends to very large distances in every direction except along $y < 0$. The dimension of the footing along the normal to the plane of

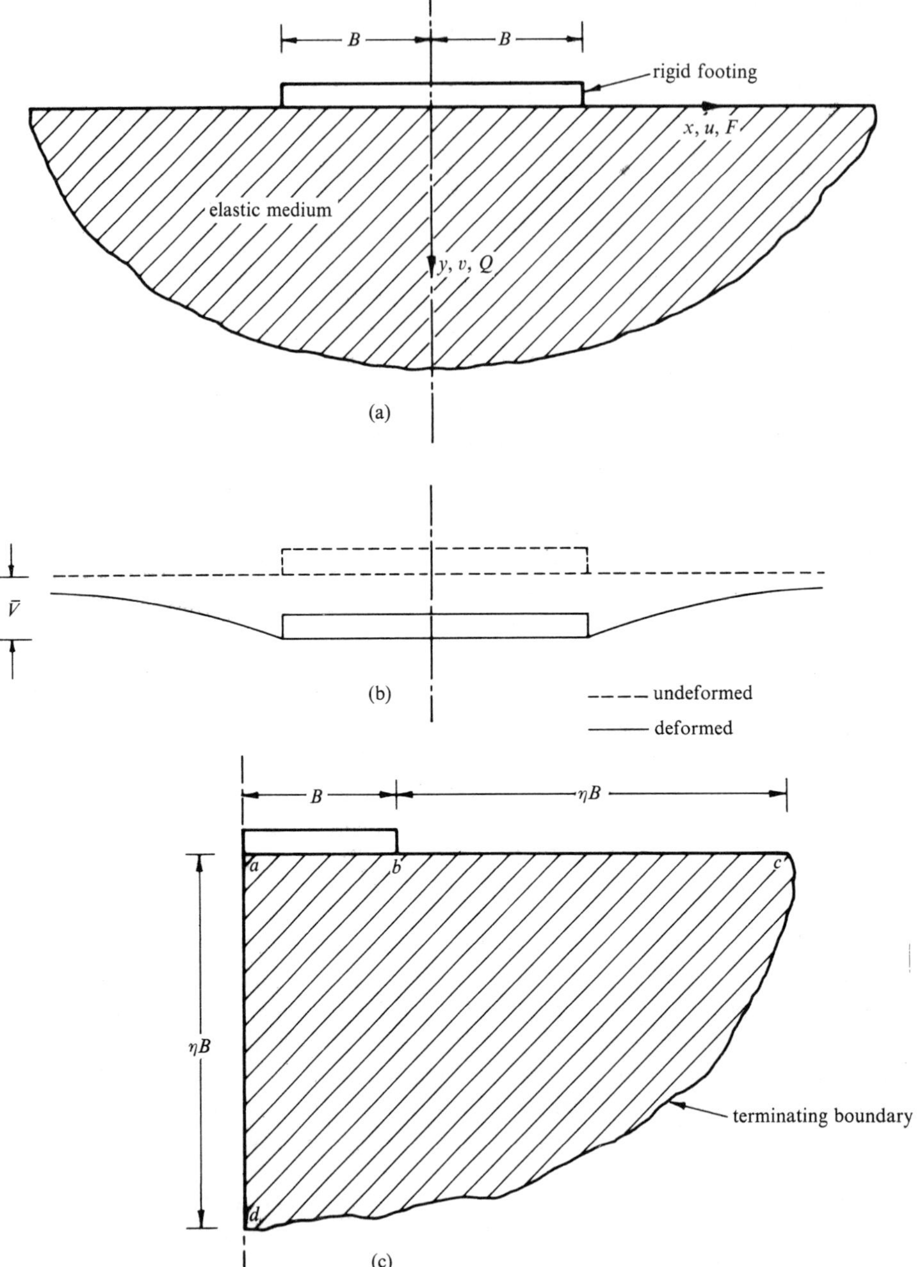

Fig. 11.7 (a) A rigid strip footing, resting on top of an isotropic and homogeneous elastic medium. (b) Uniform vertical indentation of the footing. (c) Finite approximation to an infinite medium.

paper is large compared with its width $2B$, and the common 'interface' between the footing and the medium is completely frictionless. The footing is subjected to externally applied forces such that it indents uniformly into the medium by the amount $\bar{V}$, as shown in Fig. 11.7b. To determine the contact stress σ_{yy} at the common interface $y = 0$, $-B \leqslant x \leqslant B$.

This is a typical plane strain problem. The dimension of the footing-medium along the normal to the plane of paper will therefore be taken as unity, representing a plane strain slice. Furthermore, owing to the uniform indentation of the footing the problem is a symmetrical one. We need therefore consider any symmetrical half of its geometry, as shown in Fig. 11.7c.

Let us now subdivide the symmetrical half of Fig. 11.7c into a system of triangular elements, assuming that we do not know, a priori, even the approximate stress distribution within the medium caused by the indenting footing. We can guess, however, that the stresses will vary more rapidly in the vicinity of the footing than away from it. The subdivision can therefore be 'eased off' (section 9.6) as we move away from the footing. Moreover, although the medium extends to very large (infinite) distances along $x > 0$ and $y > 0$, we must nevertheless replace the infinite medium by a finite one. This step is obviously necessary for restricting the theoretically infinite size of the overall stiffness matrix of the medium to a finite and manageable one. Let us terminate the domain of interest by a 'terminating boundary', shown in Fig. 11.7c, the extent of which is determined by the parameters η and η'.

The determination of the values of η and η' is an important aspect of analysis here since the accuracy of solution in this type of problem improves as η and η' are increased. Ideally, they should be as large as possible; we cannot, however, increase their values indefinitely, since the size of the overall stiffness matrix in two- and three-dimensional problems of this type is roughly proportional to the square and cube of these parameters respectively.

Consider the plot of Fig. 11.8 which shows the variation of computed results with η $(= \eta')$ in a plane problem of this type. Clearly, there is a certain value of η, equal to η_0, beyond which an increase in the value of η is not accompanied by a corresponding significant improvement in accuracy. Therefore, if we accept the level of error at $\eta = \eta_0$ as being 'tolerable', then the size of the overall stiffness matrix corresponding to η_0 will be optimum. It is pointed out, however, that the manner in which results converge to their exact values depends as much, if not more, on the way the subdivision of the domain is carried out. Thus, the discretization process in a problem of this type entails the determination of the optimum values of parameters like η, together with an intelligent scheme of subdivision such that a maximum degree of accuracy is achieved with the minimum possible size of the overall stiffness matrix. Finite element modelling of this type is essentially an acquired skill which improves with practice, and an experienced analyst is often able to proceed merely by inspection.

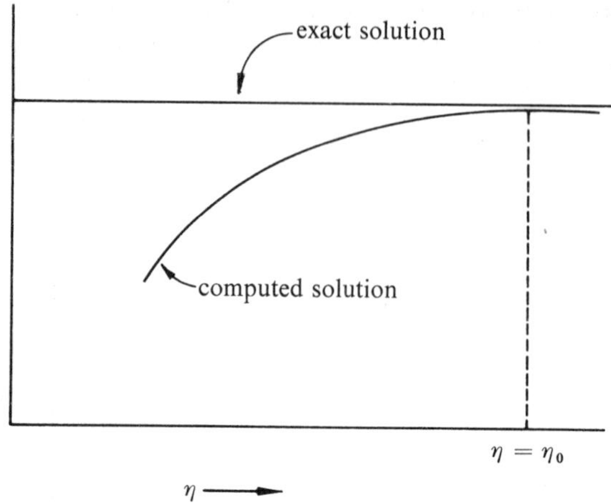

Fig. 11.8 Convergence of computed solution.

The overall stiffness matrix. Figure 11.9 shows part of a preliminary (to be subsequently modified if necessary) subdivision of the medium into a system of triangular elements with $\eta = \eta' = 5$. (In this problem all displacements occur exclusively on the x–y plane; moreover, since the medium is isotropic, the local axes of individual elements will be assumed to be parallel to the *yox* axes. Consequently, the question of transformation (section 11.4) does not arise here). The total number of nodes here is 72, of which the nodes 1–5 represent the interface ab (Fig. 11.7c). The $[K_i]$'s of individual elements are now calculated; these are then assembled by the methods of Chapter 3 to form the

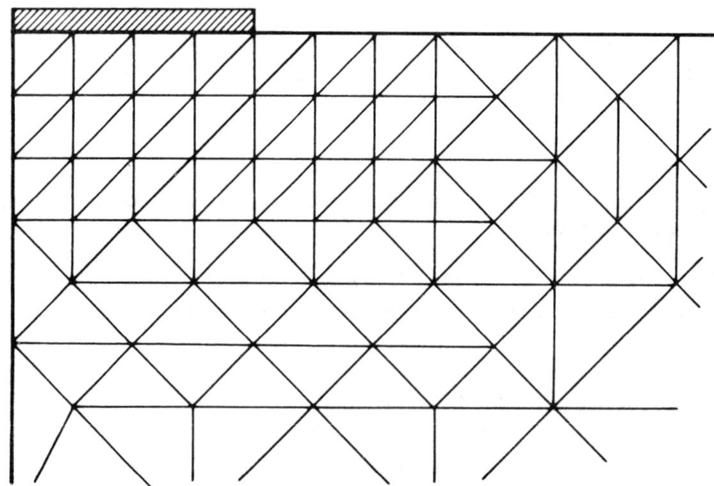

Fig. 11.9 Part of a finite element model of the medium of Fig. 11.7(c) with $\eta = \eta' = 5$.

overall stiffness matrix $[K]$ for the whole medium. In practice, these operations will be performed by the computer. A simple program for this will be given in section 15.5. For clarity of presentation, let us now write Eq. (3.9) for this problem by grouping together separately the vertical and horizontal forces and displacements as follows:

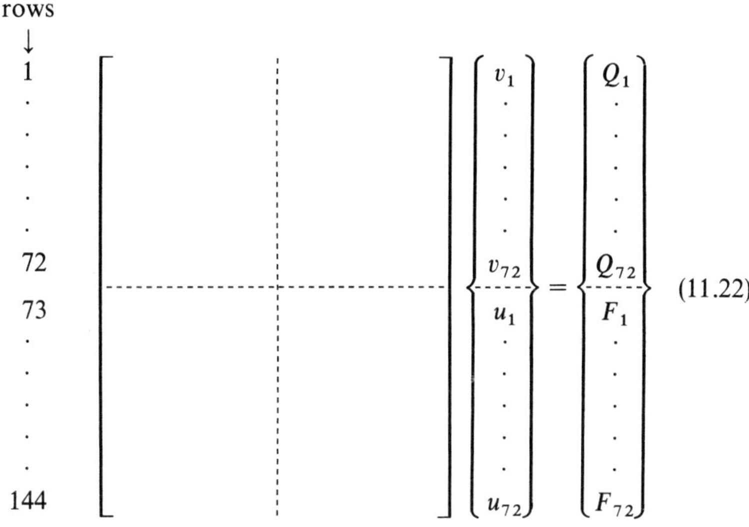

or, symbolically,

$$[K_s]\{\delta_s\} = \{P_s\} \tag{11.23}$$

Clearly, the square matrix of Eq. (11.22) which we have denoted by $[K_s]$, is obtained merely by rearranging the coefficients of $[K]$ according to the scheme of Eq. (11.22). $\{\delta_s\}$ and $\{P_s\}$ are obviously the left and right hand vectors of Eq. (11.22) respectively.

Boundary conditions. The boundary conditions of this problem and their treatment are as follows:

(a) *At the interface ab.* Since the rigid footing indents uniformly into the medium and since the nodes 1–5 represent this interface, we have

$$v_k = \bar{V} \qquad k = 1, 2, \ldots 5 \tag{11.24}$$

This condition is accounted for (method 1, section 5.1) by: (1) replacing the diagonal terms in rows 1–5 by unity and all off-diagonal terms in these rows by zero, and (2) setting each of the first five elements of $\{P_s\}$ equal to $\bar{V}$, i.e.,

$$Q_k = \bar{V} \qquad k = 1, 2 \cdots 5 \tag{11.25}$$

The other condition at this interface is concerned with forces and displacements in the x direction. Since the interface is totally frictionless,

$$\tau_{xy} = 0 \quad \text{on} \quad ab \tag{11.26}$$

The resultant of τ_{xy} is represented by the horizontal force F. Thus, by virtue of Eq. (11.26),

$$F_k = 0 \qquad k = 1, 2, \ldots 5 \tag{11.27}$$

(b) *The terminating boundary*. We shall assume that both components of displacement at the terminating boundary, *dc* (Fig. 11.7c), are negligibly small; this is valid as *dc* is sufficiently remote from the footing. Then, according to method 2 of section 5.1, the nodes representing this boundary will be excluded from Eq. (11.22).

(c) *The free surface*. In absence of externally applied forces, the following conditions prevail at the free surface, *bc* (Fig. 11.7c):

$$\sigma_{yy} = \tau_{xy} = 0 \tag{11.28}$$

Noting that the forces F and Q respectively represent the resultants of τ_{xy} and σ_{yy}, it is obvious that the condition of Eq. (11.28) will be accounted for by setting

$$F_k = Q_k = 0 \tag{11.29}$$

for any node k lying on the free surface.

(d) *The boundary due to symmetry*. As the deformation of the medium is symmetrical about $x = 0$, the horizontal displacements of any two symmetrically located points within the medium, one on each symmetrical half, will be equal in magnitude but opposite in sense. As a result, for any node k lying on $x = 0$,

$$u_k = 0 \tag{11.30}$$

This condition is accounted for by excluding from Eq. (11.22) all rows and columns which refer to such nodes (see method 2, section 5.1).

It will be observed that the 'displacement' conditions, typified by Eq. (11.24), modify the properties of the overall stiffness matrix. The 'force' or 'stress resultant' conditions typified by Eq. (11.27), on the other hand, merely contribute to the overall force vector and do not in any way modify the properties of the overall stiffness matrix.

An inspection of all the boundary conditions shows that the non-zero components of $\{P_s\}$ are provided by Eq. (11.24) alone. As a result, all computed displacements and stresses will be in terms of $\overline{V}$.

Problem solution. Having incorporated the boundary conditions into Eq. (11.22), it can now be solved uniquely for the nodal displacements, $\{\delta_s\}$, from which the stresses within individual elements can subsequently be determined.

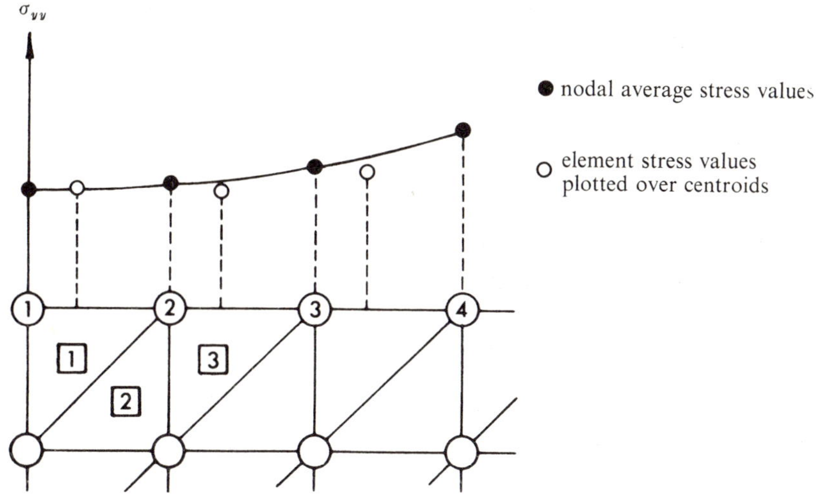

Fig. 11.10 Plotting of stress variation in triangular elements.

Since constant stresses prevail within triangular elements (section 11.1), the obvious way to depict the variation of any stress is by plotting the value of that stress in each element over the centroid of that element, as shown in Fig. 11.10. Alternatively, better results are usually obtained by taking the stress value at a node as being the average value of stresses within all elements meeting at that node. Thus, the ordinate at node 2 (Fig. 11.10), for instance, represents the value of σ_{yy} averaged between the elements 1, 2 and 3.

Curve 1 in Fig. 11.11 shows the computed variation of σ_{yy} at the contact surface, *ab*. Agreement with the corresponding exact solution[4] is seen to be reasonably good, except around $x = B$ where the theoretical value of σ_{yy} is infinity. Curve 2 represents an improved solution, based on a finer subdivision near $x = B$, in which η $(= \eta')$ was made equal to 10 and the total number of nodes representing the medium was 113. The solution can be further improved, if required, by using a still finer subdivision near the footing, particularly its edge at $x = B$.

11.6 Transverse isotropy

Suppose that the elastic medium of Fig. 11.7a were homogeneous but its elastic properties were different along perpendicular directions. Then the medium will be described as 'transversely isotropic'. Such properties are usually exhibited by stratified materials.

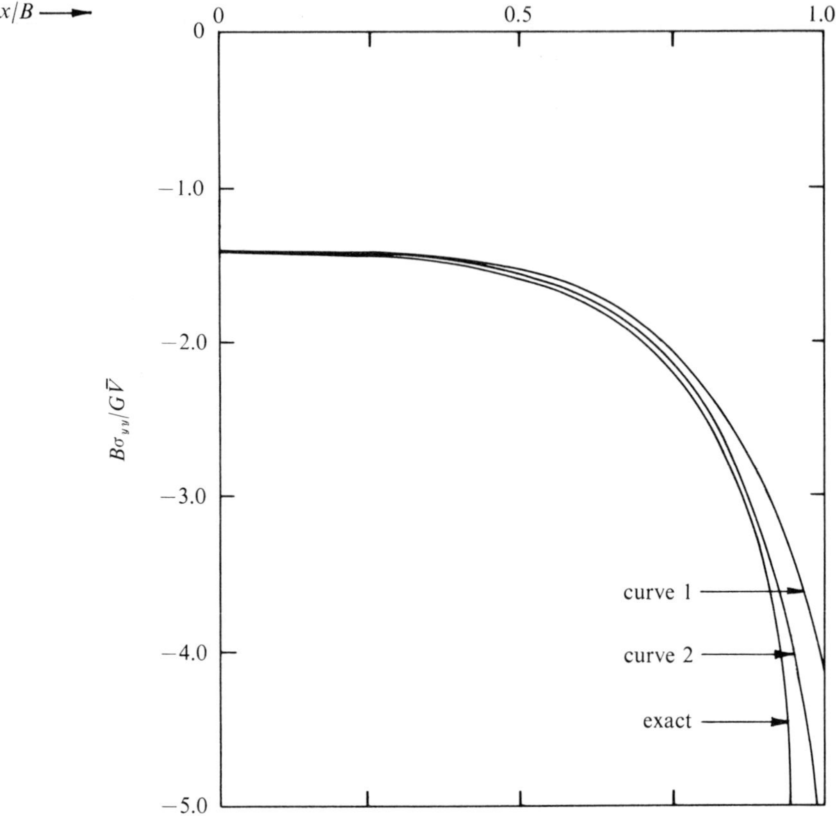

Fig. 11.11 Variation of σ_{yy} at the contact surface, assumed frictionless; $\nu = 0{\cdot}3$. G is the modulus of rigidity of the medium. Curve 1, $\eta = \eta' = 5$; curve 2, $\eta = \eta' = 10$.

Consider the plane slice of Fig. 11.12 in which stratification occurs as shown. Let E_1 and ν_1 be the elastic properties defining material behaviour along x and let E_2, ν_2 and G_2 be those along the y axis.* Then, according to Lekhnitskii,[5] the elasticity matrices will be as follows:

Plane strain

$$[d] = \begin{bmatrix} A_{11} & A_{13} & 0 \\ A_{13} & A_{33} & 0 \\ 0 & 0 & A_{55} \end{bmatrix} \tag{11.31}$$

* $E_1 = 2(1 + \nu_1)G_1$ but $E_2 \neq 2(1 + \nu_2)G_2$.

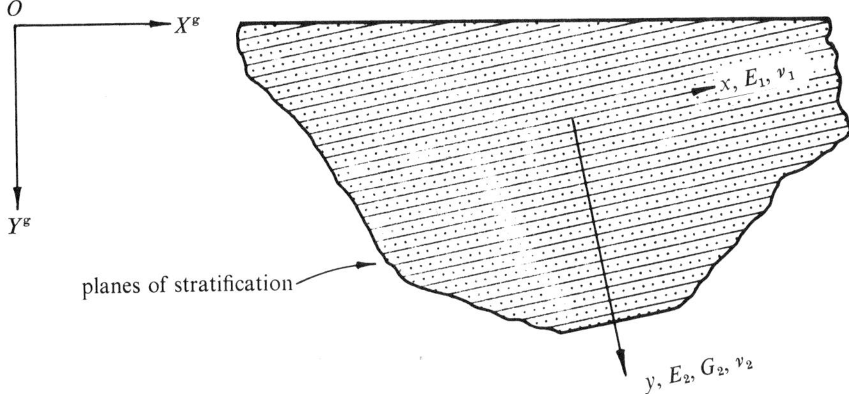

Fig. 11.12 A transversely isotropic material; stratification occurs along planes normal to the y axis.

in which

$$A_{11} = nE_2(1 - n\nu_2^2)/q$$
$$A_{13} = nE_2\nu_2(1 + \nu_1)/q$$
$$A_{33} = E_2(1 - \nu_1^2)/q$$
$$A_{55} = mE_2$$

where,

$$q = (1 + \nu_1)(1 - \nu_1 - 2n\nu_2^2), \qquad n = E_1/E_2 \quad \text{and} \quad m = G_2/E_2$$

Plane stress

The [d] matrix is given by Eq. (11.31) where now

$$A_{11} = nE_2/q,$$
$$A_{13} = nE_2\nu_2/q,$$
$$A_{33} = E_2/q$$
$$A_{55} = mE_2$$

in which $q = (1 - n\nu_2^2)$.

In transversely isotropic problems the material behaviour is clearly determined by the five mutually independent elastic constants E_1 (or G_1), ν_1, E_2, G_2 and ν_2. These problems can be solved by finite elements merely by using the above [d] matrices in place of those of Chapter 8. Note, however, that if, as in Fig. 11.12, the axes of transverse isotropy do not coincide with the global axes, then appropriate transformations (section 11.4) must be carried out.

References

1. Clough, R. W., 'The Finite Element in plane stress analysis', Proc. 2nd A.S.C.E. Conf. in Electronic Computation, Pittsburgh, Pa., Sept. 1960.

2. Turner, M. J., Clough, R. W., Martin, H. C. and Topp, L. J., 'Stiffness and deflection analysis of complex structures', *J. Aero. Sci.*, **23,** 805–23, 1956.

3. Taig, I. C., 'Structural Analysis by the displacement method', English Electric Aviation Ltd., Warton, England, Report SO 17, 1961.

4. Awojabi, A. O. and Grootenhuis, P., 'Vibration of rigid bodies on semi-infinite elastic media', *Proc. Roy. Soc.* A, **287,** 27–63, 1965.

5. Lekhnitskii, S. G., *Theory of elasticity of an anisotropic elastic body*, Holden Day, San Francisco, 1963.

12 Three dimensional stress analysis

The exact solution of three-dimensional elasticity problems by integrating the equations of Chapter 8 is nearly always an impossible proposition, as remarked in section 8.11. On the other hand, the finite element method can be used very effectively for an accurate solution of these problems, regardless of the complexity of the geometry of their domains. The body is now subdivided into a system of three-dimensional finite elements, whose characteristics can also be derived from the unit displacement theorem of Chapter 9. Of all possible element shapes, the tetrahedron and the rectangular prism are most commonly used.

12.1 The tetrahedral element

Figure 12.1 shows the extremity forces and displacements of a typical tetrahedron element. Since this element has four extremities, the variation of each component of displacement within it can be assumed in terms of a polynomial, defined by four constants* as follows:

$$u(x, y, z) = A_0 + A_1x + A_2y + A_3z \tag{12.1a}$$

$$v(x, y, z) = A_4 + A_5x + A_6y + A_7z \tag{12.1b}$$

and

$$w(x, y, z) = A_8 + A_9x + A_{10}y + A_{11}z \tag{12.1c}$$

Here A_0, A_4 and A_8 represent the rigid-body translations of the element along the x, y and z axes respectively (section 9.8(c)). Writing these equations in the matrix form of Eq. (11.1a), we have

$$\{U\} = \begin{Bmatrix} u(x, y, z) \\ v(x, y, z) \\ w(x, y, z) \end{Bmatrix}$$

$$\{A\} = \{A_0 \quad A_1 \quad \cdots \quad A_{10} \quad A_{11}\}^{\mathrm{T}}$$

and

$$[M] = \begin{bmatrix} 1 & x & y & z & 0 & 0 & 0 & 0 & 0 & 0 & 0 & 0 \\ 0 & 0 & 0 & 0 & 1 & x & y & z & 0 & 0 & 0 & 0 \\ 0 & 0 & 0 & 0 & 0 & 0 & 0 & 0 & 1 & x & y & z \end{bmatrix} \tag{12.2}$$

The extremity displacements, u_{ai}, v_{ai} etc. are now obtained by substituting the coordinates of extremities into Eqs. (12.1). Thus,

$$u_{ri} = u(x_r, y_r, z_r) \qquad v_{ri} = v(x_r, y_r, z_r)$$

and

$$w_{ri} = w(x_r, y_r, z_r)$$

* Section 9.5(a).

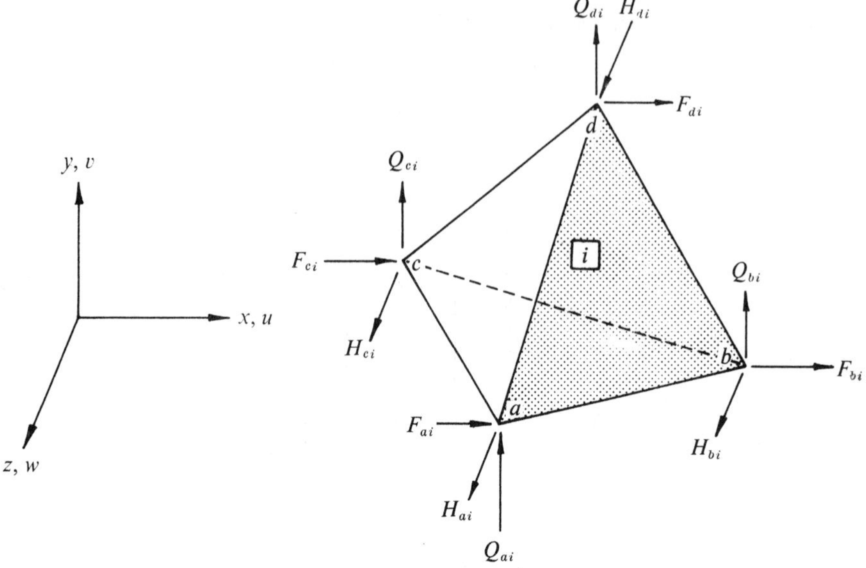

Fig. 12.1 A typical tetrahedral element (only extremity forces have been shown).

where $r = a, b, c, d$ and x_r, y_r and z_r are the coordinates of extremity r. As usual, we shall write these displacements in the following matrix form:

$$\{\bar{\delta}\} = [c]\{A\} \tag{12.3}$$

where $\{\bar{\delta}\} = \{u_{ai} \quad v_{ai} \quad w_{ai} \quad \cdots \quad u_{di} \quad v_{di} \quad w_{di}\}^{\mathrm{T}}$, and

$$[c] = \left[\begin{array}{cccc|cccc|cccc}
1 & x_a & y_a & z_a & 0 & 0 & 0 & 0 & 0 & 0 & 0 & 0 \\
0 & 0 & 0 & 0 & 1 & x_a & y_a & z_a & 0 & 0 & 0 & 0 \\
0 & 0 & 0 & 0 & 0 & 0 & 0 & 0 & 1 & x_a & y_a & z_a \\
\hline
1 & x_b & y_b & z_b & 0 & 0 & 0 & 0 & 0 & 0 & 0 & 0 \\
0 & 0 & 0 & 0 & 1 & x_b & y_b & z_b & 0 & 0 & 0 & 0 \\
0 & 0 & 0 & 0 & 0 & 0 & 0 & 0 & 1 & x_b & y_b & z_b \\
\hline
1 & x_c & y_c & z_c & 0 & 0 & 0 & 0 & 0 & 0 & 0 & 0 \\
0 & 0 & 0 & 0 & 1 & x_c & y_c & z_c & 0 & 0 & 0 & 0 \\
0 & 0 & 0 & 0 & 0 & 0 & 0 & 0 & 1 & x_c & y_c & z_c \\
\hline
1 & x_d & y_d & z_d & 0 & 0 & 0 & 0 & 0 & 0 & 0 & 0 \\
0 & 0 & 0 & 0 & 1 & x_d & y_d & z_d & 0 & 0 & 0 & 0 \\
0 & 0 & 0 & 0 & 0 & 0 & 0 & 0 & 1 & x_d & y_d & z_d
\end{array}\right]$$

The strains within the element are (Eqs. 8.1, 8.3 and 8.4)

$$\{\epsilon\} = \begin{Bmatrix} e_{xx} \\ e_{yy} \\ e_{zz} \\ e_{xy} \\ e_{yz} \\ e_{zx} \end{Bmatrix} = \begin{Bmatrix} \dfrac{\partial u}{\partial x} \\ \dfrac{\partial v}{\partial y} \\ \dfrac{\partial w}{\partial z} \\ \dfrac{\partial u}{\partial y} + \dfrac{\partial v}{\partial x} \\ \dfrac{\partial v}{\partial z} + \dfrac{\partial w}{\partial y} \\ \dfrac{\partial w}{\partial x} + \dfrac{\partial u}{\partial z} \end{Bmatrix} \tag{12.4}$$

Now, by differentiating Eqs. (12.1) according to the scheme of Eq. (12.4), we can demonstrate from Eq. (11.4b) that

$$[N] = \begin{bmatrix} 0 & 1 & 0 & 0 & 0 & 0 & 0 & 0 & 0 & 0 & 0 & 0 \\ 0 & 0 & 0 & 0 & 0 & 0 & 1 & 0 & 0 & 0 & 0 & 0 \\ 0 & 0 & 0 & 0 & 0 & 0 & 0 & 0 & 0 & 0 & 0 & 1 \\ 0 & 0 & 1 & 0 & 0 & 1 & 0 & 0 & 0 & 0 & 0 & 0 \\ 0 & 0 & 0 & 0 & 0 & 0 & 0 & 1 & 0 & 0 & 1 & 0 \\ 0 & 0 & 0 & 1 & 0 & 0 & 0 & 0 & 0 & 1 & 0 & 0 \end{bmatrix}$$

Having thus determined the explicit forms of the primary matrices $[c]$, $[M]$ and $[N]$, we can now derive the various characteristics of the element from the formulae of section 9.8. The element stiffness matrix, for instance, will now be given by

$$[K_i] = [c^{-1}]^{\mathrm{T}}[\bar{K}][c^{-1}] \tag{12.5}$$

where

$$[\bar{K}] = \int [N]^{\mathrm{T}}[d][N]\,\mathrm{d}x\,\mathrm{d}y\,\mathrm{d}z \tag{12.6}$$

in which the integration is performed throughout the entire volume of the element. Noting that $[N]$ here is independent of the current coordinates and that the following integral, when taken over the entire volume of the element, leads to

$$\int \mathrm{d}x\,\mathrm{d}y\,\mathrm{d}z = V = \text{Volume of the tetrahedron}$$

we can write $$[\bar{K}] = V \,.\, [N]^{\mathrm{T}}[d][N]$$

Obviously, the $[d]$ matrix to be used in Eq. (12.6) is that given by Eq. (8.12). The overall stiffness matrix for the whole body is now assembled, as usual, by the methods of Chapter 3, and the problem solved in terms of Eq. (3.9). The listings of $\{P\}$ and $\{\delta\}$ here are as follows:

$$\{P\} = \{F_1 Q_1 H_1 \quad F_2 Q_2 H_2 \quad \cdots \quad F_N Q_N H_N\}^{\mathrm{T}}$$

and $$\{\delta\} = \{u_1 v_1 w_1 \quad u_2 v_2 w_2 \quad \cdots \quad u_N v_N w_N\}^{\mathrm{T}}$$

where N is the total number of nodes in the entire discretized body and, at node k ($= 1, 2, \ldots, N$)

F_k, Q_k, H_k = external forces along the x, y and z directions respectively at k,

and

u_k, v_k, w_k = displacements of k along x, y and z directions respectively.

The stresses within the element will be given by Eq. (8.15), when we substitute into it

$$\{\epsilon\} = [N][c^{-1}]\begin{Bmatrix} u_n \\ v_n \\ w_n \\ u_k \\ v_k \\ w_k \\ u_m \\ v_m \\ w_m \\ u_j \\ v_j \\ w_j \end{Bmatrix}$$

in which the nodes n, k, m and j are assumed to be attached to the extremities a, b, c and d respectively.

Since $[N]$ is not a function of position, the strains and hence the stresses within the element remain constant throughout its volume. Consequently, the stress equilibrium equations are satisfied identically. We can also demonstrate, by following the procedure of Chapter 11, that the assumed displacements of Eqs. (12.1) are continuous across interfaces between adjacent elements. There is

a discontinuity of slopes across interfaces, however, as each element has a different level of strain. These properties of the tetrahedron are in fact identical to those of the triangular element of section 11.1.

If the body force accelerations are, or can be assumed to be constant throughout the entire volume of the element, then, following the procedure of section 11.1 we can demonstrate that a quarter of the total body force will be sustained by each node attached to the element. This simple result is of considerable practical importance, and can be derived by noting that the integrals

$$\int x \, dx \, dy \, dz, \qquad \int y \, dx \, dy \, dz \quad \text{and} \quad \int z \, dx \, dy \, dz$$

vanish when carried out throughout the entire volume of the element, provided that the centroid of the element is taken as the origin of coordinates.

12.2 The rectangular prism element

As the rectangular prism element of Fig. 12.2 has eight extremities, we can assume each displacement component to vary within it as a polynomial defined by eight constants.* Thus, let

$$u(x', y', z') = A_0 + A_1x' + A_2y' + A_3z' + A_4x'y' + A_5y'z' + A_6z'x' + A_7x'y'z' \tag{12.7a}$$

$$v(x', y', z') = A_8 + A_9x' + A_{10}y' + A_{11}z' + A_{12}x'y' + A_{13}y'z' + A_{14}z'x' + A_{15}x'y'z' \tag{12.7b}$$

and

$$w(x', y', z') = A_{16} + A_{17}x' + A_{18}y' + A_{19}z' + A_{20}x'y' + A_{21}y'z' + A_{22}z'x' + A_{23}x'y'z' \tag{12.7c}$$

in which the non-dimensional coordinates, $0x'y'z'$, are related to the $0xyz$ system through

$$\begin{Bmatrix} x \\ y \\ z \end{Bmatrix} = \begin{bmatrix} h_x & 0 & 0 \\ 0 & h_y & 0 \\ 0 & 0 & h_z \end{bmatrix} \begin{Bmatrix} x' \\ y' \\ z' \end{Bmatrix} \tag{12.8}$$

Here again, we have chosen the functions of Eqs. (12.7) such that displacements vary linearly at the faces of the element. The displacement on any face of the element can then be shown to depend only on those of the nodes (or extremities) defining that particular face and, as a result, is continuous across interfaces

* Section 9.5(a).

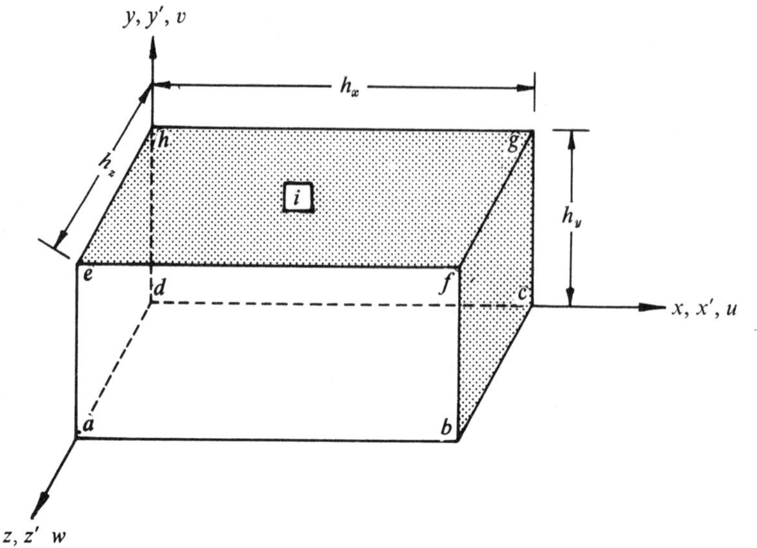

Fig. 12.2 The rectangular prism ('brick') element.

between adjacent rectangular prism elements. For this element, clearly

$$[M] = \begin{bmatrix} L & 0 & 0 \\ 0 & L & 0 \\ 0 & 0 & L \end{bmatrix}$$

in which $\{L\} = \{1 \quad x' \quad y' \quad z' \quad y'z' \quad z'x' \quad z'x' \quad x'y'z'\}$

The $[c]$ matrix is now determined by evaluating from Eqs. (12.7) the extremity displacements according to

$$u_{ri} = u(x'_r, y'_r, z'_r)$$
$$v_{ri} = v(x'_r, y'_r, z'_r)$$

and
$$w_{ri} = w(x'_r, y'_r, z'_r)$$

where $r = a, b, c, d, e, f, g, h$ and x'_r, y'_r and z'_r are the non-dimensional coordinates of extremity r. In this case $[c]$ will obviously be a 24×24 matrix.

The $[N]$ matrix will be found by first differentiating Eqs. (12.7) according to the scheme of Eq. (12.4), and then expressing the result in terms of Eq. (11.4b) (observe that the strains will have to be transformed according to Eq. (12.8); this is not necessary if the $0xyz$ coordinates are used to define the displacements). Once the primary matrices have been determined explicitly, the various steps leading to a solution will be exactly the same as those with tetrahedral elements.

A differentiation of Eqs. (12.7) shows that the strains, and hence the stresses

within the element vary in a linear fashion. The assumed displacements of Eqs. (12.7) therefore violate the stress equilibrium equations (Eq. 8.21).

12.3 Concluding remarks

Compared to the tetrahedron, the rectangular prism element has a limited use owing to its inability to approximate curved or awkward boundaries satisfactorily. The tetrahedron, on the other hand, has more flexibility in this respect and is generally preferred for this reason. In any case, the rectangular prism or other possible element shapes can be regarded as combinations of tetrahedra.

Clearly, the primary matrices of three-dimensional elements, particularly the rectangular prism, are considerably larger in size than their plane counterparts. From the point of view of automatic computation, however, this does not present any difficulty, although the computer storage requirement is now considerably greater. In three-dimensional elements, as indeed in two-dimensional elements, the $[c]$ matrix usually has zeros on its leading diagonal, and therefore cannot be directly inverted. This difficulty is easily overcome by the methods given in section 15.3.

13 Elasto-dynamic and field problems

13.1 Elasto-dynamic problems

In the foregoing chapters the forces acting on the elastic body were always assumed to be static, i.e., time-independent. However, if these forces vary with time, then the problem becomes a 'dynamic' one. Unlike its static counterpart an elasto-dynamic problem is essentially one in which 'stress waves', initiated by the exciting forces, propagate within the body which deforms as a result in a time-dependent fashion. Thus, the analysis of these problems essentially entails a study of the manifestations of wave motion within the body.

The finite element method can be used very effectively to analyze elasto-dynamic problems. However, whereas the finite element solution of a static problem is characterized by the determination of its stiffness matrix only, that of a dynamic one entails the determination of the following additional matrices:

The 'mass' matrix. The 'inertia' of the body represents its inertial resistance to vibration. In the discretized form, the total inertia of the body will be represented by what we shall call its 'overall mass matrix'.

The 'damping' matrix. In 'damped' vibration, resistance is also provided by the internal friction of the body which results in energy dissipation. The extent and nature of energy dissipation in this way is primarily determined by the damping properties of the material of the body. The total damping of the body will be represented by its 'overall damping matrix'. In addition to internal friction, energy is also dissipated by so called 'radiation damping' which occurs in problems whose domain is infinite; this will be discussed in detail in section 13.2.8.

Elasto-dynamic problems frequently occur in science and engineering. From the design point of view, the need for an accurate solution of these problems cannot be over emphasized. In this section we shall consider the application of finite elements to some fundamental aspects of elasto-dynamics.

13.1.1 *The element 'mass' matrix*

Equation (8.21a) can also be written as

$$\frac{\partial \sigma_{xx}}{\partial x} + \frac{\partial \tau_{xy}}{\partial y} + \frac{\partial \tau_{xz}}{\partial z} + \rho(\bar{X} + \bar{X}_t) = 0 \qquad (13.1.1)$$

in which
$$\bar{X}_t = -\frac{\partial^2 u}{\partial t^2} \tag{13.1.2a}$$

and t denotes time.

An inspection of Eq. (13.1.1) shows that if we treat $\bar{X}_t$ as the 'dynamic body force acceleration' along x, then $\rho(\bar{X} + \bar{X}_t)$ will obviously represent the sum of static and dynamic body forces (per unit volume) along x. Substituting

$$\bar{Y}_t = -\frac{\partial^2 v}{\partial t^2} \tag{13.1.2b}$$

and
$$\bar{Z}_t = -\frac{\partial^2 w}{\partial t^2} \tag{13.1.2c}$$

in Eqs. (8.21b) and (8.21c) respectively and expressing these equations in the form of Eq. (13.1.1), we can therefore make the following observation:

If the dynamic nature of the problem is represented by the 'dynamic body force accelerations' $\bar{X}_t$, $\bar{Y}_t$ and $\bar{Z}_t$, then the dynamic response of the body (in excess of its static response due to $\bar{X}$, $\bar{Y}$ and $\bar{Z}$) can be evaluated by imagining the body to be subject to the dynamic body forces resulting from $\bar{X}_t$, $\bar{Y}_t$ and $\bar{Z}_t$.

Then, following the procedure used for deriving Eq. (9.19), we can demonstrate that the general expression for the 'dynamic body force vector' for any finite element i is

$$\{P_{\text{b}}\}_{\text{dynamic}} = [c^{-1}]^{\text{T}} \int_V \rho [M]^{\text{T}} \begin{Bmatrix} \bar{X}_t \\ \bar{Y}_t \\ \bar{Z}_t \end{Bmatrix} dv \tag{13.1.3}$$

In a vibrating body u, v, w, $\bar{X}_t$, $\bar{Y}_t$, $\bar{Z}_t$ and $\{P_{\text{b}}\}_{\text{dynamic}}$ are all functions of both space and time; let these quantities vary with time in a 'simple harmonic'* manner such that the following representations are valid:

$$\begin{Bmatrix} u \\ v \\ w \end{Bmatrix} = \begin{Bmatrix} u_0 \\ v_0 \\ w_0 \end{Bmatrix} \sin(\omega t) \tag{13.1.4}$$

and
$$\{P_{\text{b}}\}_{\text{dynamic}} = \{P_{\text{B}}\} \sin(\omega t) \tag{13.1.5}$$

where ω is the 'circular frequency' of vibration and u_0, v_0, w_0 and $\{P_{\text{B}}\}$ are the 'steady state amplitudes' of u, v, w and $\{P_{\text{b}}\}_{\text{dynamic}}$ respectively. Then, from

* In general, this assumption is a physically valid one.

Eqs. (13.1.2)–(13.1.5) we can easily show that

$$\{P_B\} = \omega^2[c^{-1}]^T \int_V \rho[M]^T \begin{Bmatrix} u_0 \\ v_0 \\ w_0 \end{Bmatrix} dv \tag{13.1.6}$$

But, in the case of a dynamic problem, we have by definition (section 9.8)

$$\{U\} = \begin{Bmatrix} u_0 \\ v_0 \\ w_0 \end{Bmatrix} = [M]\{A\} = [M][c^{-1}]\{\delta\} \tag{13.1.7}$$

and consequently,

$$\{P_B\} = \omega^2[c^{-1}]^T \int_V \rho[M]^T[M]\, dv[c^{-1}]\{\delta\} \tag{13.1.8}$$

or

$$\{P_B\} = \omega^2[m_i]\{\delta\} \tag{13.1.9}$$

in which

$$[m_i] = [c^{-1}]^T \int_V \rho[M]^T[M]\, dv[c^{-1}] \tag{13.1.10}$$

$[m_i]$ is called the 'mass' matrix of element i. The structure of Eq. (13.1.10) shows that $[m_i]$ is a square symmetric matrix, its size being equal to that of $[K_i]$. Observe that both $[K_i]$ and $[m_i]$ of a given finite element are derived from the same assumed displacement function. For this reason $[m_i]$ is said to be 'consistent' with $[K_i]$, and is also known as the 'consistent mass matrix'[1] of element i.

To simplify subsequent analysis, we shall partition $[m_i]$ into the submatrices $\overline{m}(i, J, K)$'s, in exactly the same way as we partitioned $[K_i]$ into $S(i, J, K)$'s. Thus, if element i has n extremities, then its $[m_i]$ will be partitioned into a total of n^2 square submatrices of equal size, and these submatrices will be denoted by $\overline{m}(i, J, K)$.

The following examples illustrate the application of Eq. (13.1.10).

Example 1. Consider the bar element shown in Fig. 13.1, which has a single axial degree of freedom only. Then, as in Eq. (9.20), we have

$$\begin{aligned} U = u_0(x) &= A_0 + A_1 x \\ &= \{1 \quad x\}\{A\} \end{aligned}$$

where $\{A\} = \{A_0 \quad A_1\}^T$. Therefore, from Eq. (13.1.7)

$$\{M\} = \{1 \quad x\} \tag{13.1.11}$$

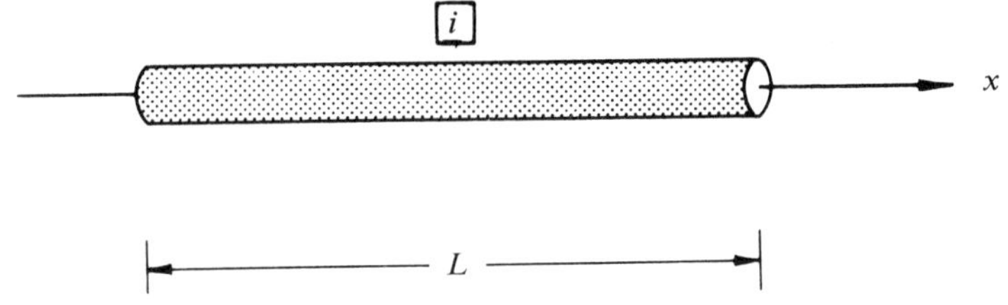

Fig. 13.1 A uniform bar element of cross section a.

Also, we have from Eq. (9.28)

$$[c^{-1}] = \begin{bmatrix} 1 & 1 \\ -\dfrac{1}{L} & \dfrac{1}{L} \end{bmatrix} \tag{13.1.12}$$

From Eq. (13.1.11) we can show that

$$[X] = \int_V \rho\{M\}^{\mathrm{T}}\{M\}\,\mathrm{d}v = \rho a \int_0^L \{1 \quad x\}^{\mathrm{T}}\{1 \quad x\}\,\mathrm{d}x = \rho a \begin{bmatrix} L & \dfrac{L^2}{2} \\ \dfrac{L^2}{2} & \dfrac{L^3}{3} \end{bmatrix}$$

Then, using $[c^{-1}]$ of Eq. (13.1.12) in

$$[m_i] = [c^{-1}]^{\mathrm{T}}[X][c^{-1}] \tag{13.1.13}$$

and forming the product of Eq. (13.1.13), we can show that the mass matrix of the bar element is

$$[m_i] = \gamma \begin{bmatrix} \frac{1}{3} & \frac{1}{6} \\ \frac{1}{6} & \frac{1}{3} \end{bmatrix} \tag{13.1.14}$$

in which $\gamma = \rho a L$ is the total mass of the element.

Example 2. Next, consider the beam element discussed in section 9.7. Writing Eq. (9.32) as

$$U = v_0(x) = \{1 \quad x \quad x^2 \quad x^3\}\{A\}$$

where $\{A\} = \{A_0 \quad A_1 \quad A_2 \quad A_3\}^{\mathrm{T}}$, we have by definition (Eq. 13.1.7)

$$\{M\} = \{1 \quad x \quad x^2 \quad x^3\} \tag{13.1.15}$$

Consequently, writing $h_i = L$ for convenience

$$[X] = \int_V \rho\{M\}^{\mathrm{T}}\{M\}\,\mathrm{d}v = \rho a\int_0^L \begin{Bmatrix} 1 \\ x \\ x^2 \\ x^3 \end{Bmatrix}\{1 \quad x \quad x^2 \quad x^3\}\,\mathrm{d}x$$

$$= \rho a\int_0^L \begin{bmatrix} 1 & & & \\ x & x^2 & \text{symm.} & \\ x^2 & x^3 & x^4 & \\ x^3 & x^4 & x^5 & x^6 \end{bmatrix}\mathrm{d}x$$

$$= \rho a\begin{bmatrix} L & & \text{symm.} & \\ \dfrac{L^2}{2} & \dfrac{L^3}{3} & & \\ \dfrac{L^3}{3} & \dfrac{L^4}{4} & \dfrac{L^5}{5} & \\ \dfrac{L^4}{4} & \dfrac{L^5}{5} & \dfrac{L^6}{6} & \dfrac{L^7}{7} \end{bmatrix}$$

Then, substituting above $[X]$ and $[c^{-1}]$ of Eq. (9.35) into Eq. (13.1.13), we can show by multiplication that for the beam element

$$[m_i] = \frac{\gamma}{420}\begin{bmatrix} 156 & & \text{symm.} & \\ 22L & 4L^2 & & \\ 54 & 13L & 156 & \\ -13L & -3L^2 & -22L & 4L^2 \end{bmatrix} \tag{13.1.16}$$

where $\gamma = \rho a L$ is the total mass of the element. Clearly, the submatrices of this $[m_i]$ are

$$\overline{m}(i, 1, 1) = \frac{\gamma}{420}\begin{bmatrix} 156 & 22L \\ 22L & 4L^2 \end{bmatrix} \tag{13.1.17a}$$

$$\overline{m}(i, 1, 2) = \frac{\gamma}{420}\begin{bmatrix} 54 & -13L \\ 13L & -3L^2 \end{bmatrix} \tag{13.1.17b}$$

$$\overline{m}(i, 2, 2) = \frac{\gamma}{420}\begin{bmatrix} 156 & -22L \\ -22L & 4L^2 \end{bmatrix} \tag{13.1.17c}$$

and of course $\overline{m}(i, 1, 2)$ and $\overline{m}(i, 2, 1)$ are transposes of each other.

Example 3. As a final example, let us derive the mass matrix of the plane stress element shown in Fig. 13.2. Using $[M]$ of Eq. (11.1b) and the integrals of Eqs. (A3.8), we can show that

$$[X] = \int_V \rho[M]^{\mathrm{T}}[M]\,\mathrm{d}v = \rho p\int_\Delta \begin{bmatrix} 1 & 0 \\ x & 0 \\ y & 0 \\ 0 & 1 \\ 0 & x \\ 0 & y \end{bmatrix} \begin{bmatrix} 1 & x & y & 0 & 0 & 0 \\ 0 & 0 & 0 & 1 & x & y \end{bmatrix} \mathrm{d}x\,\mathrm{d}y$$

$$= \rho p\int_\Delta \begin{bmatrix} 1 & & & & & \\ x & x^2 & & \text{symm.} & & \\ y & xy & y^2 & & & \\ 0 & 0 & 0 & 1 & & \\ 0 & 0 & 0 & x & x^2 & \\ 0 & 0 & 0 & y & xy & y^2 \end{bmatrix} \mathrm{d}x\,\mathrm{d}y$$

$$= \rho p\,\Delta \begin{bmatrix} 1 & & & & & \\ 0 & \frac{L^2}{18} & & & \text{symm.} & \\ 0 & -\frac{LH}{36} & \frac{H^2}{18} & & & \\ 0 & 0 & 0 & 1 & & \\ 0 & 0 & 0 & 0 & \frac{L^2}{18} & \\ 0 & 0 & 0 & 0 & -\frac{LH}{36} & \frac{H^2}{18} \end{bmatrix}$$

It is clear from Fig. 13.2 that in this case

$$x_a = -\frac{L}{3}, \quad x_b = \frac{2L}{3}, \quad x_c = -\frac{L}{3}, \quad y_a = -\frac{H}{3}, \quad y_b = -\frac{H}{3} \quad \text{and} \quad y_c = \frac{2H}{3}$$

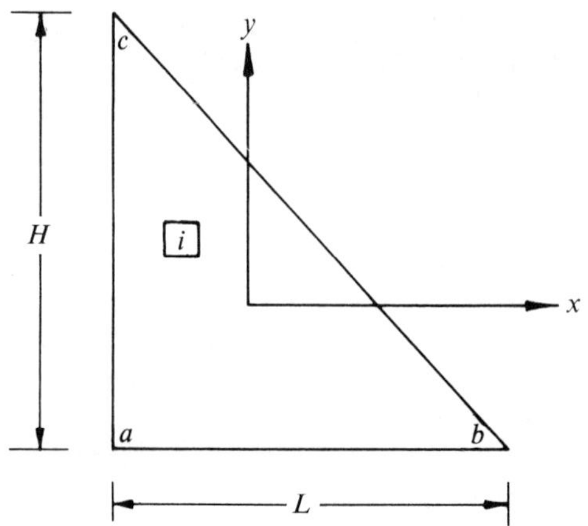

Fig. 13.2 A triangular plane element of constant thickness p (origin at centroid).

Substituting these coordinates in Table 11.1, we obtain

$$[c^{-1}] = \begin{bmatrix} \frac{1}{3} & 0 & \frac{1}{3} & 0 & \frac{1}{3} & 0 \\ -\frac{1}{L} & 0 & \frac{1}{L} & 0 & 0 & 0 \\ -\frac{1}{H} & 0 & 0 & 0 & \frac{1}{H} & 0 \\ 0 & \frac{1}{3} & 0 & \frac{1}{3} & 0 & \frac{1}{3} \\ 0 & -\frac{1}{L} & 0 & \frac{1}{L} & 0 & 0 \\ 0 & -\frac{1}{H} & 0 & 0 & 0 & \frac{1}{H} \end{bmatrix}$$

It is a simple matter now to show by forming the matrix product of Eq. (13.1.13) that

$$[m_i] = \frac{\gamma}{3}\begin{bmatrix} \frac{1}{2} & & & \text{symm.} & & \\ 0 & \frac{1}{2} & & & & \\ \frac{1}{4} & 0 & \frac{1}{2} & & & \\ 0 & \frac{1}{4} & 0 & \frac{1}{2} & & \\ \frac{1}{4} & 0 & \frac{1}{4} & 0 & \frac{1}{2} & \\ 0 & \frac{1}{4} & 0 & \frac{1}{4} & 0 & \frac{1}{2} \end{bmatrix} \qquad (13.1.18)$$

in which $\gamma = \rho p \, \Delta$ is the total mass of the element. With $[M]$ defined by Eq. (11.1b), it can be verified that $[m_i]$ of Eq. (13.1.18) will obtain for a triangular element of any shape and thickness p.

13.1.2 *Undamped forced vibration*

It was pointed out in section 9.8 that the body forces of a finite element must be treated as external forces, applied at the nodes to which the element is attached. Consequently, for every element of the discretized body, its dynamic body forces will be calculated from Eq. (13.1.8) and applied as external forces at the appropriate nodes. Then, following the method of Chapter 3, we can demonstrate that the dynamic body force-displacement relationship of the entire discretized body is

$$\{P_{\mathrm{B}}\}_{\mathrm{overall}} = \omega^2[m]\{\delta\} \tag{13.1.19}$$

where vector $\{P_{\mathrm{B}}\}_{\mathrm{overall}}$, which lists all the dynamic nodal body forces, will be treated as an external force. The matrix $[m]$ will be called the 'overall mass matrix' of the entire discretized body, since it is easily verified from Chapter 3 that $[m]$ is assembled from the $\overline{m}(i, J, K)$'s in exactly the same way as $[K]$ is from the $S(i, J, K)$'s.

Now, quite apart from the body forces of Eq. (13.1.19), imagine that the body is also subjected to other time-dependent external nodal forces whose 'harmonic amplitude' is $\{P_0\}$. Then, by definition, the total external force acting on the body is

$$\{P\} = \{P_{\mathrm{B}}\}_{\mathrm{overall}} + \{P_0\} \tag{13.1.20}$$

From Eqs. (3.9), (13.1.19) and (13.1.20) it therefore follows that

$$[K - \omega^2 m]\{\delta\} = \{P_0\} \tag{13.1.21}$$

Equation (13.1.21) is in fact the equation of undamped forced motion of the discretized body. At a given frequency of vibration (i.e., given value of ω) and 'excitation' $\{P_0\}$, the solution of Eq. (13.1.21) will give the displacement 'response' $\{\delta\}$ of the body.

13.1.3 *Natural frequencies*

There are certain values of ω at which the determinant of Eq. (13.1.21) vanishes, i.e.,

$$\det |[K] - \omega^2[m]| = 0 \tag{13.1.22}$$

When this happens, irrespective of the magnitude of $\{P_0\}$, Eq. (13.1.21) will give infinite values of $\{\delta\}$. This situation is referred to as 'resonance'. Physically,

the body resonates thereby causing intolerable stresses and strains within itself, whenever the frequency of excitation becomes equal to one of the so called 'natural frequencies' of the body. By definition, the natural frequencies of a body are frequencies at which it will vibrate, without energy dissipation, when it is momentarily excited by a force which is then removed. In practical dynamical problems the determination of natural frequencies is of paramount importance, since these are in fact the frequencies at which, for the sake of its safety, the body must not be excited.

Clearly, the natural frequencies of the body are the roots of Eq. (13.1.22). However, since in practical problems $[K]$ (and hence $[m]$) usually has a large size, it is more convenient to calculate the natural frequencies as the roots of the following 'eigenvalue' equation, which results by setting $\{P_0\} = \{0\}$ in Eq. (13.1.21):

$$[K]\{\delta\} = \lambda[m]\{\delta\} \tag{13.1.23}$$

in which $\lambda = \omega^2$.

The solution of Eq. (13.1.23) will give the roots ('eigenroots') and the vectors ('eigenvectors'). If $[K]$ (and hence $[m]$) is a $n \times n$ matrix, then Eq. (13.1.23) will have n roots and n vectors. The k-th root λ_k will be equal to ω_k^2, where ω_k is the natural frequency of the body in its k-th 'mode' of vibration. Also, the k-th vector $\{\delta_k\}$ will represent the 'mode shape' of the body when it vibrates at frequency ω_k.

Premultiplying both sides of Eq. (13.1.23) by $[K^{-1}]$, the frequency equation of the body can be expressed in its more familiar form as

$$[K^{-1}m]\{\delta\} = \frac{1}{\lambda}\{\delta\} \tag{13.1.24}$$

In practical problems the low natural frequencies of the body are of main interest. Simple computer programs can be written to extract these from Eq. (13.1.24). Alternatively, a special method given in reference 2, can be used to extract all the roots and eigenvectors of Eq. (13.1.24).

The following examples illustrate the application of Eq. (13.1.22):

Example 1. Consider the uniform cantilever beam of Fig. 13.3a. In order to find its 'fundamental' (i.e., lowest natural) frequency of transverse vibration, let us represent the entire beam by a single beam element as shown in Fig. 13.3b. In this particular case the beam is not subject to any axial or torsional forces or displacements; its $[K_i]$ will therefore be given by Eq. (2.8), when we delete from this equation all rows and columns which refer to axial and torsional modes. Then, applying the method of section 3.4, we can show that since node 1 is excluded from $[K]$,

$$[K] = S(1, 2, 2) = q\begin{bmatrix} 12 & -6L \\ -6L & 4L^2 \end{bmatrix}$$

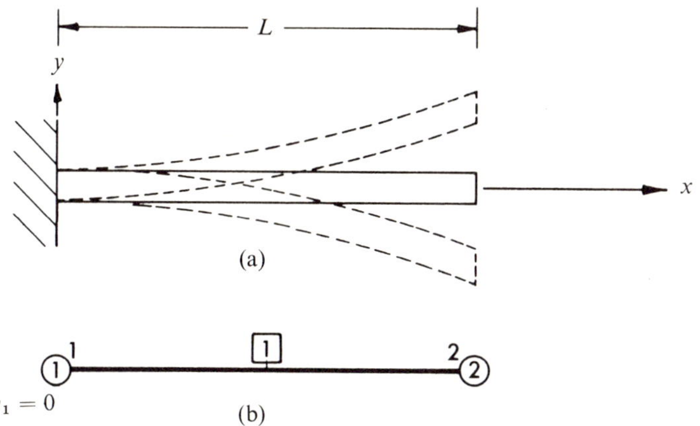

Fig. 13.3 (a) Transverse vibration of a uniform cantilever; EI constant and area of cross section = a. (b) A single element approximation of (a).

where $q = EI/L^3$. Furthermore, from Eq. (13.1.17c), the overall 'consistent mass matrix' is found to be

$$[m] = \bar{m}(1, 2, 2) = r\begin{bmatrix} 156 & -22L \\ -22L & 4L^2 \end{bmatrix}^*$$

where $r = \rho aL/420$. Then, Eq. (13.1.22) for this beam becomes

$$\det\begin{vmatrix} (12q - 156r\omega^2) & (22r\omega^2 - 6q)L \\ (22r\omega^2 - 6q)L & 4(q - r\omega^2)L^2 \end{vmatrix} = 0$$

Expansion of this determinant leads to

$$35r^2\omega^4 - 102qr\omega^2 + 3q^2 = 0$$

whose roots are

$$\omega^2 = \frac{6EI}{\rho aL^4}(102 \pm (9984)^{\frac{1}{2}})$$

or, $$\omega_1 = 3{\cdot}533(EI/\rho aL^4)^{\frac{1}{2}} \quad \text{and} \quad \omega_2 = 34{\cdot}807(EI/\rho aL^4)^{\frac{1}{2}}$$

The smaller root ω_1 is the 'fundamental' frequency of the beam; ω_2, which is the next higher root, is called its 'first harmonic' frequency. The exact[3] values of these frequencies for this beam are

$$\omega_1 = 3{\cdot}515(EI/\rho aL^4)^{\frac{1}{2}} \quad \text{and} \quad \omega_2 = 22{\cdot}034(EI/\rho aL^4)^{\frac{1}{2}}$$

It is interesting to observe that even with a crude single element approximation of the beam, the computed value of ω_1 differs from its exact value by a

* Note carefully that for a given discretized body the $[m]$ matrix will be assembled from the $\bar{m}(i, J, K)$'s in exactly the same way as $[K]$ is from the $S(i, J, K)$'s.

mere 0·51 percent. The difference between the computed and exact values of ω_2 is, however, considerable. The computed values would improve and approach their exact values, as always, as we increase the number of elements subdividing the beam. This is a general observation that applies to all bodies. It will also be observed that in general, the accuracy of computed frequencies will deteriorate with increasing mode number; that is, the computed value of ω_1 will be most accurate, that of ω_2 less accurate than ω_1 and so on.

[Suppose that instead of using the mass matrix of Eq. (13.1.16), we approached this problem from a 'common sense point of view', by imagining half the total mass of the element to be 'lumped' at each node. Then, the 'lumped mass matrix' of the beam becomes

$$[m] = \frac{\rho a L}{2}\begin{bmatrix} 1 & 0 \\ 0 & 0 \end{bmatrix}$$

Consequently, Eq. (13.1.22) now reads

$$\det\begin{vmatrix} (12q - \rho a L\omega^2/2) & -6qL \\ -6qL & 4qL^2 \end{vmatrix} = 0$$

Extracting the root of this equation, we can show that

$$\omega_1 = 2{\cdot}450(EI/\rho a L^4)^{\frac{1}{2}}$$

Clearly, this result is grossly inaccurate compared with that due to the 'consistent mass matrix'. In general, for a given subdivision of the body, the frequencies obtained by using the 'lumped mass matrix' will be far less accurate compared with those due to the 'consistent mass matrix'.]

Example 2. Since the $[m]$ matrix of a finite element refers to its 'local' coordinates, it is obvious that in the case of frames, shells etc. the $[m_i]$'s of the elements representing the body must be transformed so that they all refer to a common 'global' system of coordinates. As a typical illustration of this, consider the plane pin-jointed frame shown in Fig. 13.4a in which both members have the same area of cross section a, Young's modulus E and mass density ρ.

It will be observed that in this and similar problems the global counterparts of $[m_i]$'s will be obtained as

$$[m_i^g] = [T_i][m_i][T_i]^{\mathrm{T}} \tag{13.1.25}$$

Then, from Eqs. (7.4), (13.1.14) and (13.1.25) we can show that if the pin-jointed frame element i has length L, area of cross section a and mass density ρ, then

$$\overline{m}^g(i, 1, 1) = \overline{m}^g(i, 2, 2) = \frac{\rho a L}{3}\begin{bmatrix} l_i^2 & l_i m_i \\ l_i m_i & m_i^2 \end{bmatrix} \tag{13.1.26a}$$

and

$$\overline{m}^g(i, 1, 2) = \overline{m}^g(i, 2, 1) = \frac{\rho a L}{6}\begin{bmatrix} l_i^2 & l_i m_i \\ l_i m_i & m_i^2 \end{bmatrix} \tag{13.1.26b}$$

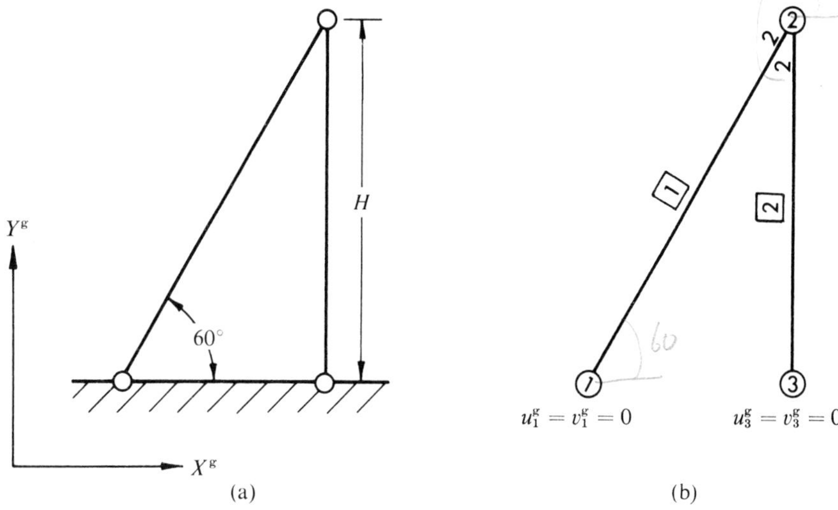

Fig. 13.4 (a) A plane pin-jointed frame. (b) Finite element model of (a).

in which $\overline{m}^g(i, J, K)$'s are the global versions of $\overline{m}(i, J, K)$'s.

Now, using the method of section 3.4, we can show that since all displacements vanish at the nodes 1 and 3 (Fig. 13.4b)

$$[K^g] = S^g(1, 2, 2) + S^g(2, 2, 2)$$

Calculation of the right hand side of this equation from Eq. (7.13) leads to

$$\begin{aligned}[K^g] &= \frac{\sqrt{3}\,aE}{8H}\begin{bmatrix} 1 & \sqrt{3} \\ \sqrt{3} & 3 \end{bmatrix} + \frac{aE}{H}\begin{bmatrix} 0 & 0 \\ 0 & 1 \end{bmatrix} \\ &= q\begin{bmatrix} 1 & \sqrt{3} \\ \sqrt{3} & 3 + \dfrac{8}{\sqrt{3}} \end{bmatrix} \end{aligned} \tag{13.1.27}$$

where $q = \sqrt{3}\,aE/8H$.

Also, we can show that the global overall mass matrix is

$$[m^g] = \overline{m}^g(1, 2, 2) + \overline{m}^g(2, 2, 2)$$

Calculation of the right hand side of this equation from Eq. (13.1.26a) gives

$$\begin{aligned}[m^g] &= \frac{\rho aH}{6\sqrt{3}}\begin{bmatrix} 1 & \sqrt{3} \\ \sqrt{3} & 3 \end{bmatrix} + \frac{\rho aH}{3}\begin{bmatrix} 0 & 0 \\ 0 & 1 \end{bmatrix} \\ &= r\begin{bmatrix} 1 & \sqrt{3} \\ \sqrt{3} & 3 + 2\sqrt{3} \end{bmatrix} \end{aligned} \tag{13.1.28}$$

where $r = \rho aH/6\sqrt{3}$.

Clearly, in the case of frames, shells etc. Eq. (13.1.22) will read

$$\det |[K^g] - \omega^2[m^g]| = 0 \tag{13.1.29}$$

Making substitutions from Eqs. (13.1.27) and (13.1.28) into Eq. (13.1.29), and expanding, it is now a simple matter to show that the natural frequencies of the frame will be the roots of the equation

$$\omega^2 = \frac{4q}{3r}$$

or

$$\omega_1 = (3E/\rho H^2)^{\frac{1}{2}}$$

which is the 'fundamental frequency' of the frame. The fact that in this case the above determinant has a single root, indicates that the frame of Fig. 13.4a has a single mode of vibration.

13.1.4 *Damped forced vibration*

The energy dissipated within the discretized body due to internal friction is usually represented by the so called 'viscous damping matrix', which we shall denote by $[C]$. The viscous damping forces set up within the body which resist motion are proportional to the particle velocities. For this reason $[C]$ is also called the 'velocity damping matrix'.

Noting that the damping forces oppose motion, it is easily verified from the theory of dynamics that the equation of damped forced motion of the discretized body is

$$[K]\{\delta\} + [C]\frac{\mathrm{d}\{\delta\}}{\mathrm{d}t} + [m]\frac{\mathrm{d}^2\{\delta\}}{\mathrm{d}t^2} = \{P\} \tag{13.1.30}$$

in which t denotes time and both $\{\delta\}$ and $\{P\}$ are time-dependent.

Owing to the presence of the velocity damping term, the response $\{\delta\}$ of the body will now be 'complex', having two components, one 'real' and the other 'imaginary'. Let

$$\{\delta\} = \{\delta_{\text{re}}\}\sin(\omega t) + \{\delta_{\text{im}}\}\cos(\omega t)$$

and

$$\{P\} = \{P_0\}\sin(\omega t)$$

in which $\{\delta_{\text{re}}\}$ and $\{\delta_{\text{im}}\}$ respectively denote the 'real' and 'imaginary' components of the 'complex response' $\{\delta\}$. Then, substitution of $\{\delta\}$ and $\{P\}$ into Eq. (13.1.30) leads to the following two equations:

$$[K - \omega^2 m]\{\delta_{\text{re}}\} - \omega[C]\{\delta_{\text{im}}\} = \{P_0\} \tag{13.1.31a}$$

$$\omega[C]\{\delta_{\text{re}}\} + [K - \omega^2 m]\{\delta_{\text{im}}\} = \{0\} \tag{13.1.31b}$$

The $[C]$ matrix depends on the viscous damping properties of the material of the body. When it is specified, the real and imaginary components of response

at a given ω can be found by solving the above two equations simultaneously. Often, it is more convenient to express and solve these equations in the following alternative form:

$$\begin{bmatrix} [K - \omega^2 m] & -\omega[C] \\ \omega[C] & [K - \omega^2 m] \end{bmatrix} \begin{Bmatrix} \{\delta_{re}\} \\ \{\delta_{im}\} \end{Bmatrix} = \begin{Bmatrix} \{P_0\} \\ \{0\} \end{Bmatrix}$$

13.2 The 'field' equations

Numerous physical phenomena are governed by what are known as 'field' equations; of these, we shall discuss the finite element method of solution of the following which occur frequently in various scientific and engineering disciplines:

The Poisson equation,
The Laplace equation and
The wave equation.

In each case, our objective is to determine the 'field function', which we denote by ϕ, at all the nodes of the discretized domain of the problem. In addition to satisfying the particular equation that governs it, ϕ must also satisfy the prescribed boundary conditions of the problem.

It is not necessary at this stage to specify exactly what ϕ represents; in fact, it could represent any physical quantity whose values within the domain of the problem are governed by any of the above equations.

The characteristics of finite elements to deal with these equations are most satisfactorily derived from the methods of calculus of variations. However, as this requires a relatively high level of mathematical skill, we shall derive them here from a simple alternative method of 'elastic analogy', in which we shall think of the domain of the problem as an 'elastic' continuum and establish an analogy between the 'field' and 'elastic' quantities. Apart from its inherent simplicity, the main advantage of this approach lies in the fact that the treatment of a field problem now reduces to that of an analogous 'elastic' problem, to be solved in terms of Eq. (3.9).

An example of the variational approach will be given in Appendix 2.

13.2.1 *The Poisson equation*

To begin with, it would be instructive to focus our attention on a specific, simple physical phenomenon which is governed by this equation. Consider, for example, the flow of electric current in a purely resistive two-dimensional conductor, a differential element of which is shown in Fig. 13.5; here, in addition to the current intensities (defined as current/area) i_x and i_y entering *ABCD*, a current density (defined as current/volume) of $I(x, y)$ is also introduced

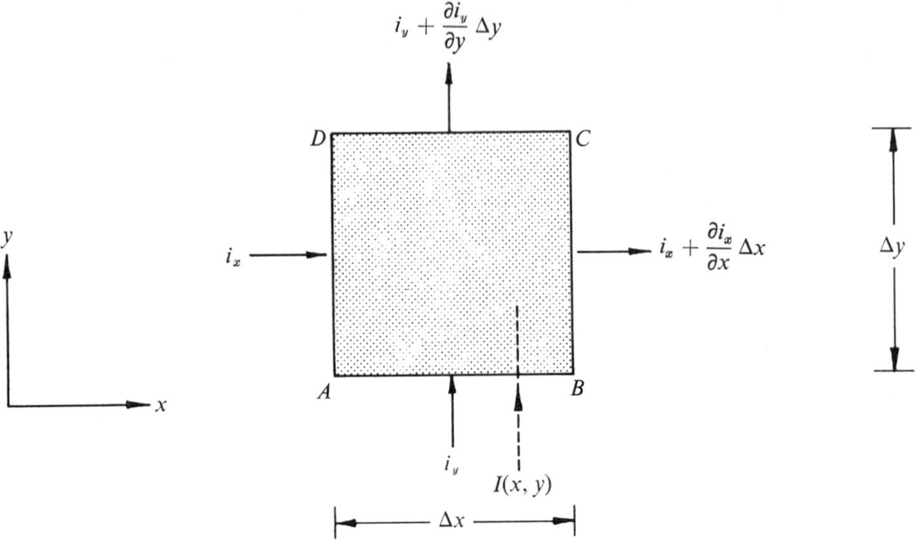

Fig. 13.5 A differential element of a two-dimensional resistive conductor.

into *ABCD* by means of a number of 'sources', one of which is shown by the dotted line. It is clear from this diagram that:

Total current entering $ABCD = i_x\,\Delta y + i_y\,\Delta x + I(x, y)\,\Delta x\,\Delta y$* and total current leaving *ABCD*

$$= \left(i_x + \frac{\partial i_x}{\partial x}\Delta x\right)\Delta y + \left(i_y + \frac{\partial i_y}{\partial y}\Delta y\right)\Delta x$$

Therefore, the total amount of current accumulated by *ABCD*

$$\left(\frac{\partial i_x}{\partial x} + \frac{\partial i_y}{\partial y} - I(x, y)\right)\Delta x\,\Delta y$$

However, since a conductor is not able to accumulate electric current, we obtain by setting the above expression equal to zero

$$\frac{\partial i_x}{\partial x} + \frac{\partial i_y}{\partial y} = I(x, y) \qquad (13.2.1)$$

Equation (13.2.1), known as the 'continuity condition', ensures the continuity of current flow within the conductor and may be regarded as the electrical counterpart of displacement compatibility of section 3.1.

Now, according to Ohm's law the current intensities are related to the potential

* We have assumed here that $I(x, y)$ does not vary within *ABCD*.

(voltage) function $\phi(x, y)$ through the equations

$$i_x = -\frac{1}{r_x} \cdot \frac{\partial \phi}{\partial x} \tag{13.2.2a}$$

and

$$i_y = -\frac{1}{r_y} \cdot \frac{\partial \phi}{\partial y} \tag{13.2.2b}$$

where r_x and r_y denote specific electrical resistances of the conductor along x and y respectively. Substituting for i_x and i_y into Eq. (13.2.1), we get

$$\frac{\partial}{\partial x}\left(\frac{1}{r_x} \cdot \frac{\partial \phi}{\partial x}\right) + \frac{\partial}{\partial y}\left(\frac{1}{r_y} \cdot \frac{\partial \phi}{\partial y}\right) = -I(x, y) \tag{13.2.3}$$

Equation (13.2.3) represents the general form of the Poisson equation in two-dimensional rectangular coordinates. We can demonstrate likewise that in a three-dimensional Cartesian space it becomes

$$\frac{\partial}{\partial x}\left(\frac{1}{r_x} \cdot \frac{\partial \phi}{\partial x}\right) + \frac{\partial}{\partial y}\left(\frac{1}{r_y} \cdot \frac{\partial \phi}{\partial y}\right) + \frac{\partial}{\partial z}\left(\frac{1}{r_z} \cdot \frac{\partial \phi}{\partial z}\right) = -I(x, y, z) \tag{13.2.4}$$

A variety of physical phenomena is governed by the Poisson equation; of these the torsion of elastic prismatic bars, transverse deflection of thin loaded membranes etc. are well known examples. All these phenomena have, among others, the following in common:

(1) A single-valued function (which we have denoted by ϕ) exists throughout the entire domain of the problem.

(2) The 'gradient variables' (which we have denoted by i_x, i_y and i_z) are related to ϕ by means of equations similar to Eqs. (13.2.2), and

(3) r_x, r_y and r_z represent some property of the domain; in anisotropic domains they represent this property along the coordinate axes, while in case of isotropy $r_x = r_y = r_z = r$.

13.2.2 *An 'elastic analogy' for the solution of the Poisson equation*

Equation (13.2.2) can be written in the matrix form as

$$\begin{Bmatrix} i_x \\ i_y \end{Bmatrix} = \begin{bmatrix} \dfrac{1}{r_x} & 0 \\ 0 & \dfrac{1}{r_y} \end{bmatrix} \begin{Bmatrix} -\dfrac{\partial \phi}{\partial x} \\ -\dfrac{\partial \phi}{\partial y} \end{Bmatrix} \tag{13.2.5}$$

When dealing with the Poisson equation, we shall think of the domain of the problem as an 'elastic body'; in two dimensions the 'stresses' and 'strains'

within the body will be defined as

$$\{\sigma\} = \begin{Bmatrix} i_x \\ i_y \end{Bmatrix} \tag{13.2.6}$$

and

$$\{\epsilon\} = \begin{Bmatrix} -\dfrac{\partial\phi}{\partial x} \\[2ex] -\dfrac{\partial\phi}{\partial y} \end{Bmatrix} \tag{13.2.7}$$

respectively. Then, comparing Eq. (13.2.5) with Eq. (8.15) (ignoring initial and thermal stresses), we observe that the two-dimensional Poisson equation can be solved in terms of an analogous 'elasticity' problem whose stress-strain relationship is specified by Eq. (13.2.5) and whose 'elasticity' matrix is

$$[d] = \begin{bmatrix} \dfrac{1}{r_x} & 0 \\[2ex] 0 & \dfrac{1}{r_y} \end{bmatrix} \tag{13.2.8}$$

To complete the analogy, we must also find the 'elastic' counterpart of $I(x, y)$. From its definition and the observations made in sections 8.7 and 9.4, it is clear that the characteristics of $I(x, y)$ are identical to those of body forces in elasticity, and therefore can be treated as such. Then, replacing the vector $\rho\begin{Bmatrix} \overline{X} \\ \overline{Y} \end{Bmatrix}$ by $I(x, y)$, we can show that by definition (section 9.8) the 'body forces' for the two-dimensional Poisson equation are

$$\{P_\mathrm{b}\} = [c^{-1}]^\mathrm{T}\int_S [M]^\mathrm{T} I(x, y)\,\mathrm{d}x\,\mathrm{d}y \tag{13.2.9}$$

In the case of the Poisson equation in three dimensions, the analogous 'elastic' quantities are easily verified to be as follows:

$$\{\sigma\} = \begin{Bmatrix} i_x \\ i_y \\ i_z \end{Bmatrix} \tag{13.2.10}$$

$$\{\epsilon\} = \begin{Bmatrix} -\dfrac{\partial\phi}{\partial x} \\[2ex] -\dfrac{\partial\phi}{\partial y} \\[2ex] -\dfrac{\partial\phi}{\partial z} \end{Bmatrix} \tag{13.2.11}$$

$$[d] = \begin{bmatrix} \frac{1}{r_x} & 0 & 0 \\ 0 & \frac{1}{r_y} & 0 \\ 0 & 0 & \frac{1}{r_z} \end{bmatrix} \tag{13.2.12}$$

and

$$\{P_b\} = [c^{-1}]^T \int_V [M]^T I(x, y, z)\, dv \tag{13.2.13}$$

The above analogy enables us to derive the properties of finite elements, to deal with the Poisson equation, from the methods of Chapter 9.

Using this analogy, the Poisson, Laplace and the wave equation will be solved in terms of analogous elasticity problems whose finite element solution has been described in Chapters 11 and 12.

13.2.3 *Solution of the Poisson equation*

Since in problems governed by this equation we are usually concerned with the determination of the nodal values of ϕ only, the degree of freedom is unity. In two-dimensions this equation will be solved by subdividing the 'elastic body' into a system of triangular and/or rectangular elements of uniform unit thickness, while in three-dimensions the body will be subdivided into tetrahedral and/or rectangular prism elements. The primary matrices $[M]$, $[c]$ and $[N]$ and the vector $\{P_b\}$ will be found as follows:

Two-dimensional elements. We consider first the plane triangular element shown in Fig. 13.6. According to section 9.5(a) the variation of $\phi(x, y)$ within this element can be assumed as

$$\phi(x, y) = A_0 + A_1 x + A_2 y \tag{13.2.14}$$

or

$$U = \phi(x, y) = \{1 \quad x \quad y\}\{A\}$$

where

$$\{A\} = \{A_0 \quad A_1 \quad A_2\}^T$$

Therefore, by definition,

$$\{M\} = \{1 \quad x \quad y\} \tag{13.2.15}$$

Now, substituting the extremity coordinates into Eq. (13.2.14), we obtain

$$\phi_{ai} = \phi(x_a, y_a) = A_0 + A_1 x_a + A_2 y_a$$

$$\phi_{bi} = \phi(x_b, y_b) = A_0 + A_1 x_b + A_2 y_b$$

and

$$\phi_{ci} = \phi(x_c, y_c) = A_0 + A_1 x_c + A_2 y_c$$

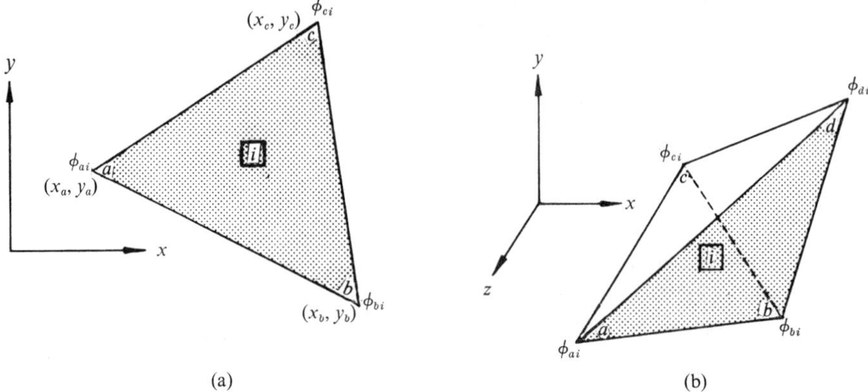

Fig. 13.6 (a) A triangular 'field' element of unit thickness. (b) A tetrahedral 'field' element.

or writing these equations in the matrix form,

$$\begin{Bmatrix} \phi_{ai} \\ \phi_{bi} \\ \phi_{ci} \end{Bmatrix} = \begin{bmatrix} 1 & x_a & y_a \\ 1 & x_b & y_b \\ 1 & x_c & y_c \end{bmatrix} \begin{Bmatrix} A_0 \\ A_1 \\ A_2 \end{Bmatrix} \tag{13.2.16}$$

or using symbolic notation

$$\{\bar{\delta}\} = [c]\{A\} \tag{13.2.17}$$

in which $[c]$ denotes the square matrix of Eq. (13.2.16). By a direct inversion of $[c]$ we can show that

$$[c^{-1}] = \frac{1}{2\Delta} \begin{bmatrix} (x_b y_c - x_c y_b) & (x_c y_a - x_a y_c) & (x_a y_b - x_b y_a) \\ (y_b - y_c) & (y_c - y_a) & (y_a - y_b) \\ (x_c - x_b) & (x_a - x_c) & (x_b - x_a) \end{bmatrix} \tag{13.2.18}$$

in which Δ = area of the triangle.

Now, differentiating Eq. (13.2.14) we have

$$-\frac{\partial \phi}{\partial x} = -A_1$$

and

$$-\frac{\partial \phi}{\partial y} = -A_2$$

or using matrix notation

$$\begin{Bmatrix} -\dfrac{\partial \phi}{\partial x} \\ -\dfrac{\partial \phi}{\partial y} \end{Bmatrix} = \begin{bmatrix} 0 & -1 & 0 \\ 0 & 0 & -1 \end{bmatrix} \begin{Bmatrix} A_0 \\ A_1 \\ A_2 \end{Bmatrix} \tag{13.2.19}$$

or, using symbolic notation,

$$\{\epsilon\} = [N]\{A\} \tag{13.2.20}$$

in which $[N]$ denotes the rectangular matrix of Eq. (13.2.19).

The calculation of $\{P_b\}$ for the triangular element is considerably simplified when its centroid is taken as the origin of coordinates, since in that case the integral

$$\int_\Delta f^n \, dx \, dy \qquad n \text{ odd}, \quad f = x, y$$

taken over the entire area of the triangle vanishes.* Then, from Eqs. (13.2.9), (13.2.15) and (13.2.18) we can demonstrate that when $I(x, y)$ has (or can be assumed to have) the constant value of I throughout the element,

$$\{P_b\} = \frac{I}{2}\begin{Bmatrix}(x_by_c - x_cy_b)\\(x_cy_a - x_ay_c)\\(x_ay_b - x_by_a)\end{Bmatrix} \tag{13.2.21}$$

(The similarity between Eqs. (11.13) and (13.2.21) is worth noting). A simple calculation will show that for a plane triangular element of any shape and unit uniform thickness, $\{P_b\}$ in fact reduces to

$$\{P_b\} = \frac{\Delta I}{3}\begin{Bmatrix}1\\1\\1\end{Bmatrix} \tag{13.2.22}$$

indicating, as in Eq. (11.13), that one-third of the total 'mass' ΔI of the triangle is sustained by each node attached to it. In practice, it is most convenient to subdivide the body in such a way that $I(x, y)$ does not vary significantly within individual elements; the simple expression of Eq. (13.2.22) may then be used to give accurate results. However, when it is not possible to do this and when the variation of $I(x, y)$ can be represented by means of a continuous function, $\{P_b\}$ will be found by performing the integration of Eq. (13.2.9).

Having thus determined the primary matrices, we can derive the various element characteristics from the formulae given in section 9.8. For example, as $[N]$ of this element is independent of the current coordinates, we can show that

$$[K_i] = \Delta[c^{-1}]^{\mathrm{T}}\begin{bmatrix}0 & 0 & 0\\0 & \dfrac{1}{r_x} & 0\\0 & 0 & \dfrac{1}{r_y}\end{bmatrix}[c^{-1}] \tag{13.2.23}$$

in which $[c^{-1}]$ is given by Eq. (13.2.18).

* See Appendix 3.

The primary matrices and subsequently the characteristics of the rectangular element can also be found in exactly the same way, by assuming the variation of $\phi(x, y)$ within the element as

$$\phi(x, y) = A_0 + A_1x + A_2y + A_3xy \tag{13.2.24}$$

Three-dimensional elements. Consider the tetrahedral element shown in Fig. 13.6. Since this element has four extremities, the variation of $\phi(x, y, z)$ within it will be assumed as the following polynomial defined by four constants:

$$\phi(x, y, z) = A_0 + A_1x + A_2y + A_3z \tag{13.2.25}$$

Writing $\{A\} = \{A_0 \quad A_1 \quad A_2 \quad A_3\}^{\mathrm{T}}$, we can show that by definition

$$\{M\} = \{1 \quad x \quad y \quad z\} \tag{13.2.26}$$

Substituting the extremity coordinates into Eq. (13.2.25) and then expressing the resulting four equations in the matrix form of Eq. (13.2.17), we can also show that

$$[c] = \begin{bmatrix} 1 & x_a & y_a & z_a \\ 1 & x_b & y_b & z_b \\ 1 & x_c & y_c & z_c \\ 1 & x_d & y_d & z_d \end{bmatrix} \tag{13.2.27}$$

The primary matrix $[N]$ will be found by differentiating Eq. (13.2.25) according to the scheme of Eq. (13.2.11). Then, expressing the 'strains' in the matrix form of Eq. (13.2.20), it is easily verified that in this case

$$[N] = \begin{bmatrix} 0 & -1 & 0 & 0 \\ 0 & 0 & -1 & 0 \\ 0 & 0 & 0 & -1 \end{bmatrix} \tag{13.2.28}$$

Taking the centroid of the tetrahedron as the origin of coordinates, we can demonstrate that if $I(x, y, z)$ has (or can be assumed to have) the constant value of I throughout the element, then

$$\{P_b\} = \frac{VI}{4}\begin{Bmatrix} 1 \\ 1 \\ 1 \\ 1 \end{Bmatrix} \tag{13.2.29}$$

in which V = total volume of the tetrahedron.

The primary matrices and subsequently the characteristics of the rectangular prism element (Fig. 12.2) may be found in exactly the same way, by assuming

$\phi(x, y, z)$ to vary within the element as

$$\phi(x, y, z) = A_0 + A_1x + A_2y + A_3z + A_4xy + A_5yz + A_6zx + A_7xyz$$

Problem solution. Having calculated the $[K_i]$'s of all elements into which the 'body' is subdivided, the $S(i, J, K)$'s will be assembled by the methods of Chapter 3 to form the overall 'stiffness' matrix $[K]$ for the entire body. Since the degree of freedom is unity, in both two- and three-dimensional elements each of the $S(i, J, K)$'s will in fact be a single quantity. The Poisson equation can then be solved in terms of Eq. (3.9), which now becomes

$$[K]\{\phi\} = \{P\} \tag{13.2.30}$$

in which $\{\phi\}$ lists the unknown nodal values of ϕ. The non-zero components of $\{P\}$ will be contributed by (a) the $\{P_b\}$'s of elements and (b) the non-zero boundary conditions, when such conditions are prescribed.

However, in order to solve Eq. (13.2.30) for a set of unique values of ϕ, we must incorporate into it all the prescribed boundary conditions; this follows in the next section.

13.2.4 *Treatment of boundary conditions*

The conditions to be satisfied by ϕ at the boundaries of the domain are usually prescribed either as

$$\phi = \phi_b \quad \text{on } S \tag{13.2.31}$$

or as

$$\frac{\partial \phi}{\partial n} + qa(s) + \beta\phi = 0 \quad \text{on } S \tag{13.2.32}$$

where q and β are constants and $a(s)$ a function. Here S may either be a part or whole of the boundary of the 'body', and $\partial\phi/\partial n$ the slope of ϕ along n, which is the outward normal to S. In general, ϕ_b, $a(s)$, ϕ and $\partial\phi/\partial n$ will all vary over S.

The condition of Eq. (13.2.31) is easily dealt with by method 1 of section 5.1; according to this, ϕ_b will be treated as a prescribed 'displacement' at every node representing S.

The condition of Eq. (13.2.32), on the other hand, presents some difficulty as the elastic analogy now breaks down and recourse must therefore be taken to the methods of calculus of variations. For instance, if we write this equation as

$$-\frac{\partial \phi}{\partial n} = qa(s) + \beta\phi$$

and persist with the elastic analogy, then the right hand side clearly appears to be equal to the 'strain' along n. Consider, for example, the situation at node k in Fig. 13.7a. Since k lies on PQ, the 'force' P_k acting at k will be given by

$$P_k = -\frac{1}{r_n}\int_A^B \frac{\partial \phi}{\partial n}\,\mathrm{d}s = \frac{q}{r_n}\int_A^B a(s)\,\mathrm{d}s + \frac{\beta}{r_n}\int_A^B \phi\,\mathrm{d}s \;^{*} \tag{13.2.33}$$

* Since, Force $= \int$ (stress) ds and stress = ('elasticity')(strain).

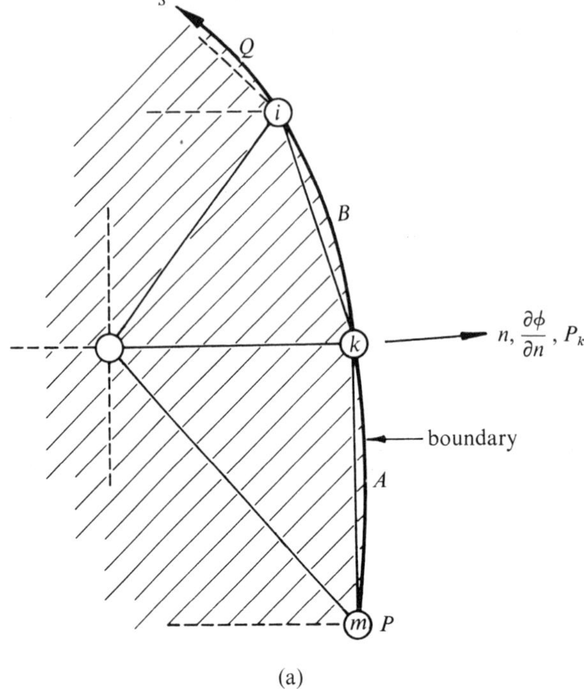

(a)

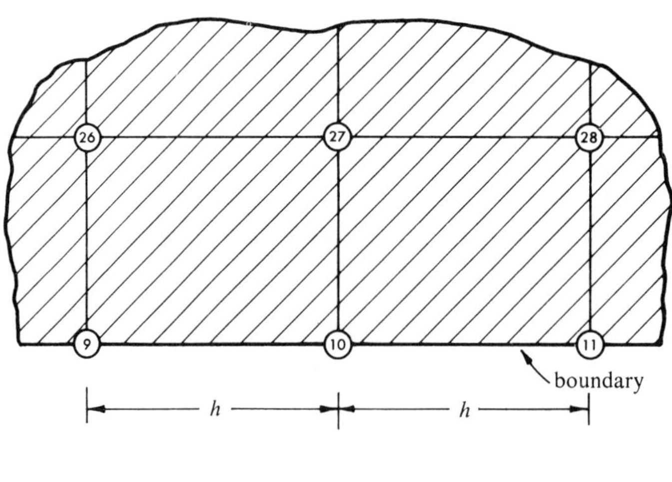

(b)

Fig. 13.7 (a) 'Gradient' boundary condition on PQ (the body has unit thickness). (b) Part of a domain divided into rectangular field elements. The condition of Eq. (13.2.32) is prescribed on the line 9-10-11.

in which $1/r_n$ is 'elasticity' along n, and A and B are the mid points of $m - k$ and $k - i$ respectively (Fig. 13.7a). Equation (13.2.33), however, is not correct since from the calculus of variations it can be shown[4] that the correct expression for P_k is

$$P_k = q\int_A^B a(s)\,\mathrm{d}s + \beta\int_A^B \phi\,\mathrm{d}s \tag{13.2.34}$$

Thus, in the finite element method the condition of Eq. (13.2.32) will be accounted for when external 'forces', computed from Eq. (13.2.34), are applied at all the nodes representing S. The integrals of Eq. (13.2.34) will be evaluated numerically; for instance, assuming that both $a(s)$ and ϕ vary linearly along the sides $m - k$ and $k - i$ in Fig. 13.7a, and making all such sides representing PQ equal, we can approximate P_k as

$$P_k = qha_k + \beta h\phi_k \tag{13.2.35}$$

where $m - k = k - i = h$, and a_k and ϕ_k are the values of $a(s)$ and ϕ at node k, respectively. Obviously, more sophisticated numerical integration schemes can be used for better accuracy.

As an example, let us now apply Eq. (13.2.35) to the boundary node 10 ($= k$) in a typical domain, a part of which is shown in Fig. 13.7b; here, the condition of Eq. (13.2.32) is prescribed on the boundary line represented by the nodes 9, 10 and 11. Since P_k is an external force, we can show that the 10-th row of Eq. (13.2.30) for this case will read

$$k_{10,9}\phi_9 + k_{10,10}\phi_{10} + k_{10,11}\phi_{11} + k_{10,26}\phi_{26} + k_{10,27}\phi_{27} + k_{10,28}\phi_{28} = qha_{10} + \beta h\phi_{10}$$

or,

$$k_{10,9}\phi_9 + (k_{10,10} - \beta h)\phi_{10} + k_{10,11}\phi_{11} + k_{10,26}\phi_{26} + k_{10,27}\phi_{27} + k_{10,28}\phi_{28} = qha_{10}$$

in which the $k_{i,j}$'s are coefficients of the $[K]$ matrix. Clearly, the right hand side qha_{10} is in fact the 10-th component of $\{P\}$. Also, notice how the coefficient(s) of ϕ (and hence the $[K]$ matrix itself) is modified when $\beta \neq 0$.

The reflecting boundary condition. Returning briefly to the electrical example of Fig. 13.5, we can write the following equation, similar to Eq. (13.2.2), for the normal direction n (Fig. 13.7a):

$$i_n = -\frac{1}{r_n}\cdot\frac{\partial\phi}{\partial n}$$

Now, if PQ represents a non-conducting boundary, then it is clear that no current will flow across PQ, i.e., $i_n = 0$. As a result,

$$\frac{\partial\phi}{\partial n} = 0 \quad \text{on } PQ \tag{13.2.36}$$

The above condition, which also results by setting $q = \beta = 0$ in Eq. (13.2.32), is called a 'reflecting' or 'non-conducting' boundary condition. This condition can be used very conveniently to reduce the size of problems whose domains have one or more axes of symmetry; we shall use this in the following example. Equation (13.2.34) shows that since $q = \beta = 0$, no external 'force' will be applied at node k, if k lies on such a reflecting boundary.

13.2.5 *An example of the solution of the Poisson equation*

The 'torsion function' ϕ of a solid, uniform and isotropic bar of arbitrary cross section is governed by the equation

$$\frac{\partial^2 \phi}{\partial x^2} + \frac{\partial^2 \phi}{\partial y^2} = -2G\theta \tag{13.2.37}$$

in which G and θ are modulus of rigidity and angle of twist per unit length, respectively. The boundary condition of this problem is that

$$\phi = 0$$

at the perimeter of the cross section of the bar.

Clearly, Eq. (13.2.37) is a Poisson equation in two dimensions which obtains by setting $r_x = r_y = G\theta$ and $I(x, y) = 2$ in Eq. (13.2.3).

Let us now solve Eq. (13.2.37) within the rectangular cross section shown in Fig. 13.8a. Clearly, if the axes of symmetry are treated as 'reflecting' boundaries, then we need consider any symmetrical quadrant of the section. Figure 13.8b shows a quadrant subdivided into a system of triangular elements, whose characteristics have been discussed in section 13.2.3.

Since $\phi = 0$ at the external boundaries, these need not be represented by nodes (method 2, section 5.1). Further, the nodes 4, 8, 9, 10, 11 and 12 representing the reflecting boundaries will be free from external 'forces'. Clearly therefore, none of the nodes of this problem is subject to external forces as such. Consequently, $\{P\}$ will derive solely from the body forces. Consider element 1 (Fig. 13.8b) for example; according to Eq. (13.2.22), a third of its 'mass' ΔI acts at node 1. Further, since 6 equal triangles meet at this node and since $I = 2$ here,

$$\text{total body force at node 1} = 6(2\Delta/3) = 2L^2$$

The quantity $2L^2$ will be treated as an external force acting at node 1. We can show in this way that in this problem

$$\{P\} = \frac{L^2}{3}\{6 \quad 6 \quad 6 \quad 3 \quad 6 \quad 6 \quad 6 \quad 3 \quad 3 \quad 3 \quad 3 \quad 2\}^{\mathrm{T}}$$

A simple computer program, based on the method developed in this chapter, for solving the Poisson and Laplace equations in two dimensions will be given in section 15.5. Using this, the nodal values of ϕ for this problem were found

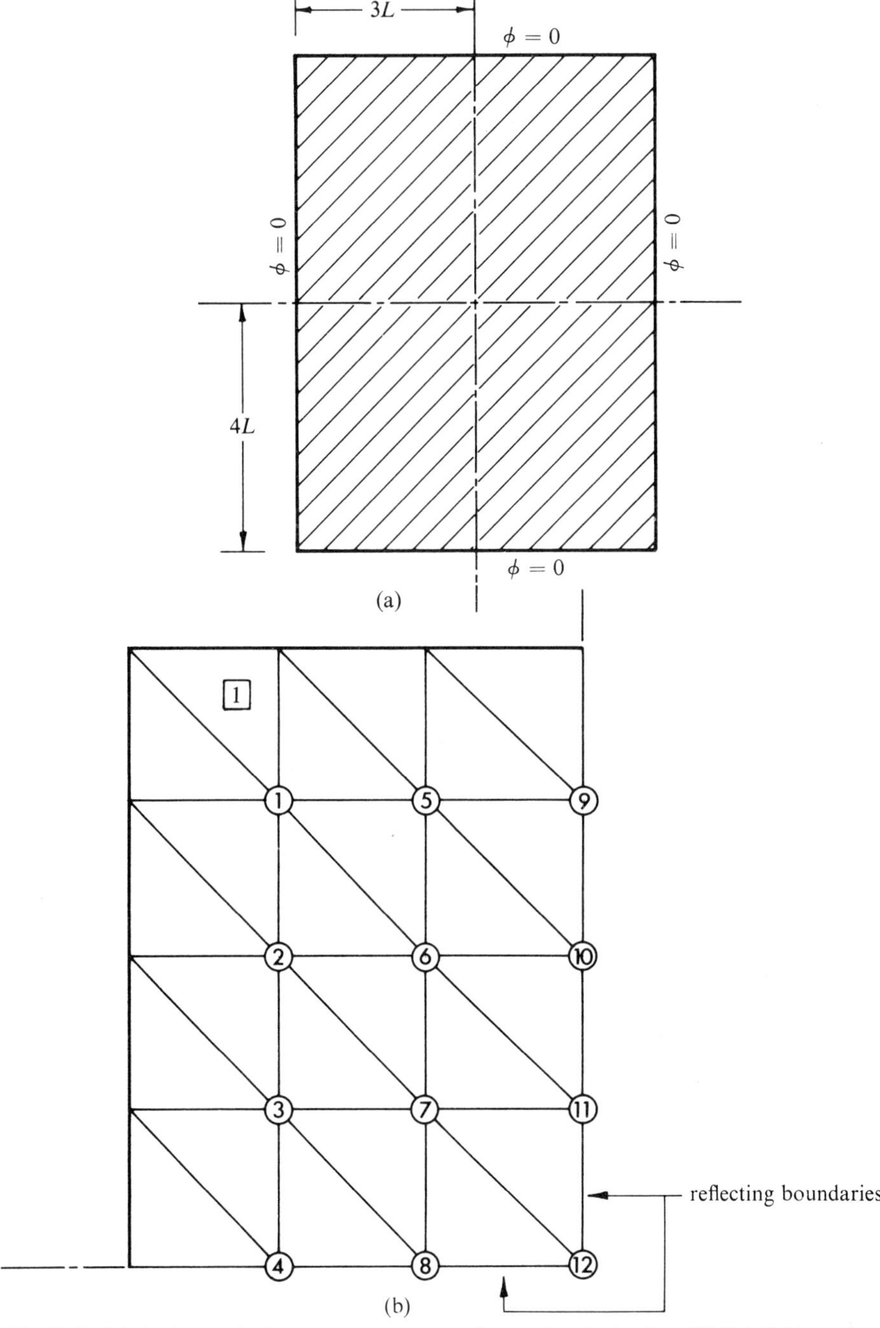

Fig. 13.8 (a) An isotropic, homogeneous rectangular section in torsion. (b) Subdivision of a quadrant of (a) into triangular elements (all triangles are isosceles).

to be as follows:

Node	Computed values of $\phi/G\theta L^2$	Exact[5] values of $\phi/G\theta L^2$
1	2·0559	2·0990
2	3·1516	3·1818
3	3·7021	3·7120
4	3·8755	3·8720
5	3·0719	3·1073
6	4.8484	4·8773
7	5·7812	5·7748
8	6·0979	6·0484
9	3·3832	3·4123
10	5·3891	5·4083
11	6·4763	6·4337
12	6·9538	6·7478

Clearly, accuracy is very good and can be further improved, if required, by using a finer subdivision.

13.2.6 *Calculation of 'stresses'*

Following the elastic analogy, the 'stresses' within individual elements of the discretized body can easily be found, once the nodal values of ϕ have been computed. For example, in the case of the tetrahedral element of Fig. 13.6b we can show that

$$\begin{Bmatrix} i_x \\ i_y \\ i_z \end{Bmatrix} = [d][Nc^{-1}]\begin{Bmatrix} \phi_n \\ \phi_k \\ \phi_m \\ \phi_j \end{Bmatrix} \tag{13.2.38}$$

in which $[d]$, $[N]$ and $[c]$ are given by Eqs. (13.2.12), (13.2.28) and (13.2.27) respectively, and the nodes n, k, m and j are assumed to be attached to the extremities a, b, c and d respectively.

13.2.7 *The Laplace equation*

When the right hand side of the Poisson equation is set equal to zero, the resulting equation is called the Laplace equation. Thus, setting $I(x, y) = 0$ in Eq. (13.2.3), the general form of the Laplace equation in two dimensions is found to be

$$\frac{\partial}{\partial x}\left(\frac{1}{r_x}\cdot\frac{\partial\phi}{\partial x}\right) + \frac{\partial}{\partial y}\left(\frac{1}{r_y}\cdot\frac{\partial\phi}{\partial y}\right) = 0 \tag{13.2.39}$$

Similarly for three dimensions.

The Laplace equation is perhaps the most important and fundamental equation of Physics. Among the numerous physical phenomena governed by it the following are well known:

Seepage in porous media,
Steady state heat flow,
Irrotational flow of ideal fluids,
Distribution of electric and magnetic potentials, etc.

Since $I(x, y) = I(x, y, z) = 0$, the 'body force' vector $\{P_b\}$ now vanishes; consequently, the non-zero components of $\{P\}$ of Eq. (13.2.30) will derive solely from the prescribed boundary conditions. Except for this difference, the finite element treatment of the Laplace equation is identical to that of the Poisson equation. A computer program for solving the Laplace equation in two dimensions will be given in section 15.5.

13.2.8 *The wave equation*

When the right hand side of the Poisson equation is replaced by

$$\rho \frac{\partial^2 \phi}{\partial t^2}$$

where t denotes time, the resulting equation is called the 'wave equation'. Thus, from Eq. (13.2.4), the general form of the wave equation in three dimensions is found to be

$$\frac{\partial}{\partial x}\left(\frac{1}{r_x} \cdot \frac{\partial \phi}{\partial x}\right) + \frac{\partial}{\partial y}\left(\frac{1}{r_y} \cdot \frac{\partial \phi}{\partial y}\right) + \frac{\partial}{\partial z}\left(\frac{1}{r_z} \cdot \frac{\partial \phi}{\partial z}\right) = \rho \frac{\partial^2 \phi}{\partial t^2} \tag{13.2.40}$$

The quantities r_x, r_y and r_z are now certain properties of the domain such that

$$1/(\rho r_x)^{\frac{1}{2}}, \qquad 1/(\rho r_y)^{\frac{1}{2}} \text{ and } 1/(\rho r_z)^{\frac{1}{2}}$$

all have dimensions of velocity. For isotropic domains $r_x = r_y = r_z = r$, and consequently Eq. (13.2.40) becomes

$$\frac{\partial^2 \phi}{\partial x^2} + \frac{\partial^2 \phi}{\partial y^2} + \frac{\partial^2 \phi}{\partial z^2} = \frac{1}{c^2} \cdot \frac{\partial^2 \phi}{\partial t^2} \tag{13.2.41}$$

in which

$$c = 1/(\rho r)^{\frac{1}{2}}$$

is called the 'wave velocity'.

A comparison shows that the wave equation, unlike the Poisson or Laplace equation, is essentially dynamic in which the 'field function' ϕ depends on both space and time. Thus, Eq. (13.2.41), for example, governs the motion of waves in an isotropic three dimensional medium in which the waves, initiated by some disturbance, travel with velocity c. A large variety of physical phenomena in which energy is transported by waves from point to point within the domain, is governed by the wave equation. Familiar examples of such

wave motion are: propagations of electromagnetic waves, gravity waves in water, stress waves in elastic bodies, etc.

The wave equation is capable of two different types of solution, namely the 'standing wave' solution and the 'progressive wave' solution. Consider, for example, the motion of water waves initiated by dropping a pebble on the still surface of a pond. Let us assume for the moment that there is no agency contributing in any way to retard the motion of these waves and that the sides of the pond are totally reflecting. In the total absence of dissipating mechanisms, it is clear that these waves travelling with their wave velocity, will be repeatedly reflected between the sides of the pond in a periodic manner, ad infinitum, giving rise to a pattern of what are called 'standing' waves. It is obvious therefore that standing waves, or rather a standing wave solution of the wave equation, will obtain whenever the domain of the wave equation has reflecting boundaries.

If, on the other hand, the boundaries of the pond were at (theoretical) infinity so that no reflection occurred, then energy will be transported away to infinity by the waves from the source of disturbance. Clearly, in the absence of reflections, there will be no 'feed back' of energy to the source of disturbance. This situation will call for what is known as a 'progressive wave' solution of the wave equation. In practice both progressive and standing waves will eventually dissipate owing to the internal friction (viscosity in the case of water) of the medium. However, in infinite domains, the transportation of energy by waves to infinity by itself constitutes an additional damping mechanism, called 'radiation damping', which is distinct from that due to internal friction. Observe that a finite domain has 'natural frequencies' which result from the repeated reflections of waves between its boundaries. In the absence of reflections an infinite domain, on the other hand, does not possess natural frequencies as such.

The reader would have realized by now that a finite element solution of the wave equation will, in general, be a standing wave solution. This is because we are not able, as a rule, to represent an infinite domain by a finite element model (recall that the question of standing or progressive waves did not arise in section 11.5 as the problem there was a static one).

Now, consider the case when ϕ is a harmonic function of time, so that in two dimensions we can write

$$\phi(x, y) = \phi' \sin(\omega t) \tag{13.2.42}$$

where ω is circular frequency of wave motion. Then, substituting for $\phi(x, y)$ into the two-dimensional version of Eq. (13.2.41), we have

$$\frac{\partial^2 \phi'}{\partial x^2} + \frac{\partial^2 \phi'}{\partial y^2} = -(\omega/c)^2 \phi' \tag{13.2.43}$$

A comparison shows that if we set $r_x = r_y = 1$, $\phi = \phi'$ and $I(x, y) = (\omega/c)^2 \phi'$ into Eq. (13.2.3), then Eq. (13.2.43) can be solved as a two-dimensional Poisson equation. Therefore, making appropriate substitution into Eq. (13.2.9), we

obtain

$$\{P_b\} = (\omega/c)^2[c^{-1}]^T \int_S [M]^T \phi' \, dx \, dy$$

or, since by definition (section 9.8) $\phi' = [M][c^{-1}]\{\bar{\delta}\}$, we have

$$\{P_b\} = (\omega/c)^2[c^{-1}]^T \int_S [M]^T[M] \, dx \, dy[c^{-1}]\{\bar{\delta}\} \tag{13.2.44}$$

or

$$\{P_b\} = (\omega/c)^2[m_i]\{\bar{\delta}\} \tag{13.2.45}$$

where

$$[m_i] = [c^{-1}]^T \int_S [M]^T[M] \, dx \, dy[c^{-1}] \tag{13.2.46}$$

(Note that $[m_i]$ above, unlike that of Eq. (13.1.10), does not contain ρ).

Comparing Eq. (13.2.46) with Eq. (13.1.10), we observe that the finite element solution of Eq. (13.2.43) will in fact proceed exactly as that of an elasto-dynamic problem discussed in section 13.1; first, the $[m_i]$'s of the individual elements subdividing the domain will be calculated from Eq. (13.2.46) (or its counterpart for three dimensions). These will then be assembled, as in the case of elasto-dynamic problems, to form the 'overall mass matrix' $[m]$ for the entire discretized domain. Then, having accounted for the prescribed boundary conditions, the unknown nodal values of ϕ' at a given frequency ω will be found by solving

$$[K - (\omega/c)^2 m]\{\phi'\} = \{P_0\} \tag{13.2.47}$$

in which $[K]$ is that of Eq. (13.2.30) and $\{P_0\}$ denotes the 'harmonic amplitude' of externally applied 'forces'. Extension of this method to the solution of the three dimensional wave equation is obvious.

As a typical application of Eq. (13.2.46), let us now calculate the $[m_i]$ matrix of the element shown in Fig. 13.9 in which the centroid of the element is taken as origin of coordinates. Assuming that

$$\phi'(x, y) = A_0 + A_1 x + A_2 y$$

we have by definition

$$\{M\} = \{1 \quad x \quad y\} \tag{13.2.48}$$

Consequently,

$$[X] = \int_\Delta \{M\}^T\{M\} \, dx \, dy = \int_\Delta \begin{Bmatrix} 1 \\ x \\ y \end{Bmatrix} \{1 \quad x \quad y\} \, dx \, dy$$

$$= \int_\Delta \begin{bmatrix} 1 & x & y \\ x & x^2 & xy \\ y & xy & y^2 \end{bmatrix} dx \, dy$$

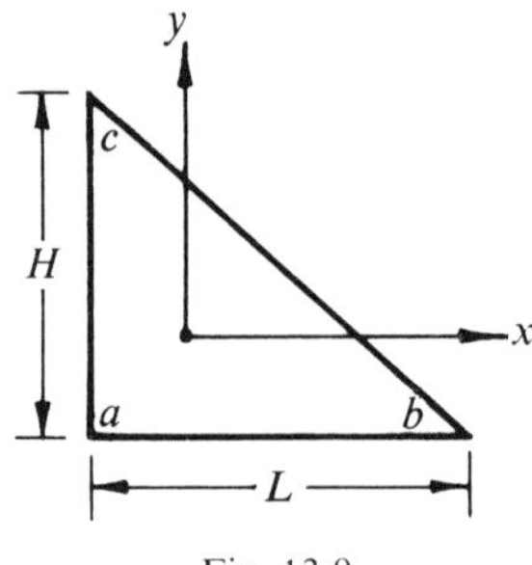

Fig. 13.9

Then, using the integrals of Eqs. (A3.8), we can show that

$$[X] = \Delta \begin{bmatrix} 1 & 0 & 0 \\ 0 & \dfrac{L^2}{18} & -\dfrac{LH}{36} \\ 0 & -\dfrac{LH}{36} & \dfrac{H^2}{18} \end{bmatrix}$$

Now, substituting the extremity coordinates of this element into Eq. (13.2.18), we have

$$[c^{-1}] = \begin{bmatrix} \dfrac{1}{3} & \dfrac{1}{3} & \dfrac{1}{3} \\ -\dfrac{1}{L} & \dfrac{1}{L} & 0 \\ -\dfrac{1}{H} & 0 & \dfrac{1}{H} \end{bmatrix}$$

Then, by forming the product of Eq. (13.2.46) we can show that

$$[m_i] = [c^{-1}]^{\mathrm{T}}[X][c^{-1}] = \frac{\Delta}{3}\begin{bmatrix} \frac{1}{2} & \frac{1}{4} & \frac{1}{4} \\ \frac{1}{4} & \frac{1}{2} & \frac{1}{4} \\ \frac{1}{4} & \frac{1}{4} & \frac{1}{2} \end{bmatrix}$$

It can be verified that with $\{M\}$ defined by Eq. (13.2.48), the above value of $[m_i]$ will obtain for a triangular element of any shape.

13.2.9 *An example of the solution of the wave equation*

Consider the two-dimensional water-filled reservoir of Fig. 13.10a in which the sides *BC* and *CD* are immovable, while side *AB* which is movable, is subjected

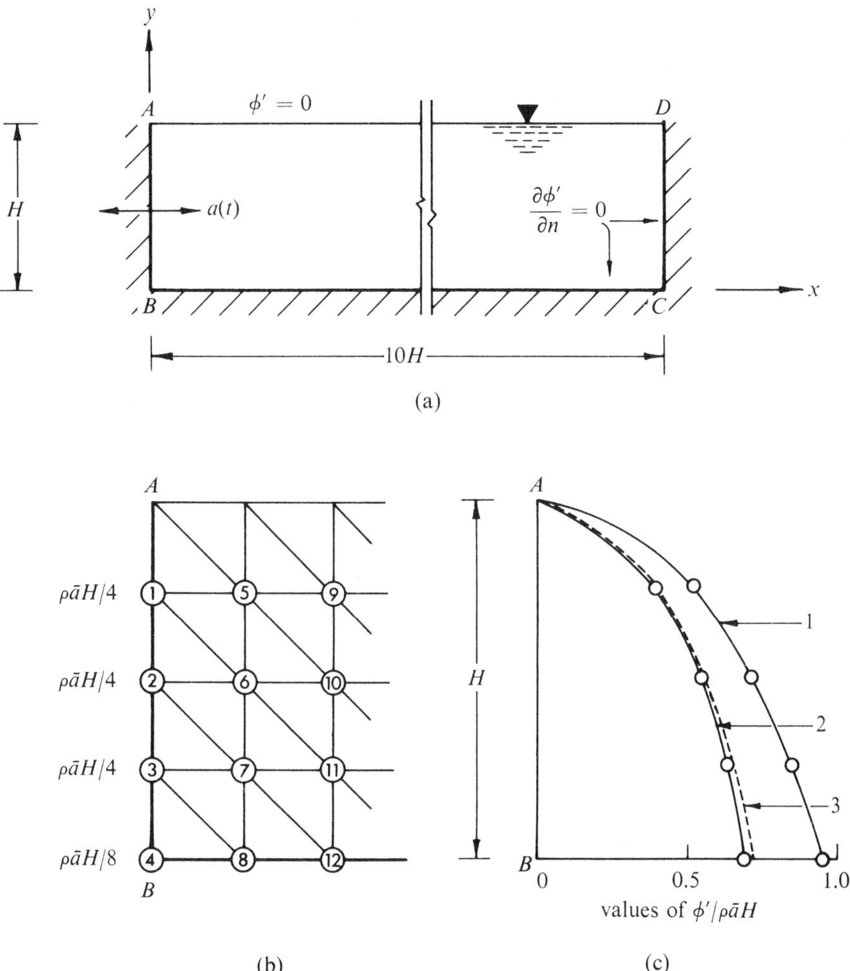

Fig. 13.10 (a) Geometry of a reservoir filled with water. (b) Finite element model of a part of (a). (c) Dynamic pressures on face AB: (1) computed, $\omega = \pi c/3H$; (2) computed, $\omega = 0$; (3) exact[6] $\omega = 0$.

to the periodic acceleration

$$a(t) = \bar{a} \sin(\omega t)$$

as shown. As a result of this motion hydrodynamic pressures $\phi(x, y, t)$ will be generated. If viscosity of water is ignored, then these pressures will vary with time as

$$\phi(x, y, t) = \phi' \sin(\omega t)$$

in which ϕ' denotes 'harmonic amplitudes' of hydrodynamic pressures. We can demonstrate[6] that for small amplitudes of motion, ϕ' will be governed by Eq. (13.2.43) in which c now denotes velocity of sound in water.

In order to find the values of ϕ' within $ABCD$, let us now subdivide it into a system of triangular elements, a part of which is shown in Fig. 13.10b. Having calculated the $[m_i]$'s from Eq. (13.2.46) and the $[K_i]$'s from the method of section 13.2.3, these are then assembled to give $[m]$ and $[K]$ respectively.

The prescribed boundary conditions of this problem are:

(a) $\phi' = 0$ at the free water surface AD; this surface need not therefore be represented by nodes.

(b) $\partial\phi'/\partial n = 0$ on BC and CD. According to Eq. (13.2.35), this 'reflecting' condition will be accounted for by applying zero external 'forces' at nodes representing these boundaries.

(c) $\partial\phi'/\partial x = -\rho\bar{a}$ on the moving boundary AB. This condition will be dealt with by Eq. (13.2.35) in which we must now set $h = H/4$, $q = \rho$, $a_k = \bar{a}$ and $\beta = 0$. Consider node 1 (Fig. 13.10b) for example. According to this equation, the external force at this node is ($k = 1$)

$$P_1 = \rho\bar{a}H/4.$$

Similarly, for the remaining nodes on this face,

$$P_2 = P_3 = \rho\bar{a}H/4 \quad \text{and} \quad P_4 = \rho\bar{a}H/8.$$

Clearly, the non-zero components of $\{P_0\}$ will therefore derive from (c) only; i.e.,

$$\{P_0\} = \rho\bar{a}H\{\tfrac{1}{4} \quad \tfrac{1}{4} \quad \tfrac{1}{4} \quad \tfrac{1}{8} \quad \cdots \quad \text{all zeros} \quad \cdots\}$$

We can now solve Eq. (13.2.47) to find the nodal values of ϕ'. Figure 13.10c shows the computed pressures on face AB at two frequencies of motion. Clearly, the accuracy of computed results can be further improved, if required, by using a finer subdivision.

It is interesting to note that the solution with $\omega = 0$ (Fig. 13.10c) is in fact the solution of the Laplace equation

$$\frac{\partial^2\phi'}{\partial x^2} + \frac{\partial^2\phi'}{\partial y^2} = 0$$

which results when we set $\omega = 0$ in Eq. (13.2.43).

References

1. Archer, J. S., 'Consistent mass matrix for distributed systems', *Proc. Amer. Soc. Civ. Eng.*, **89**, ST4, p. 161, 1963.

2. Zienkiewicz, O. C., Irons, B. and Nath. B., 'Natural frequencies of complex, free or submerged, structures by the finite element method', in *Symposium on Vibrations in Civil Engineering*, London, 1965.

3. Timoshenko, S. and Young, D. H., *Vibration problems in Engineering*, 3rd edn., D. Van Nostrand Company Inc., New York, 1955.

4. Visser, W., 'A finite element method for the determination of non-stationary temperature distribution and thermal deformations', *Proc. Conference on Matrix Methods in Structural Mechanics*, Air Force Inst. of Technology, Ohio, Oct. 1965.

5. Filonenko-Boroditch, M., *Mathematical Theory of Elasticity*, Dover Publications Inc., p. 236, 1965.

6. Westergaard, H. M., 'Water pressure on Dams during Earthquakes', *Trans. Amer. Soc. Civ. Eng.*, **98,** 1933, p. 418–33.

14 Stability problems

14.1 General

In all our discussions so far the body has been assumed to be 'geometrically linear' (section 2.1), meaning that the strains within it were related to the displacements in a linear fashion. This assumption is valid only as long as the displacements are small, so that the deformed geometry of the body is not significantly different from its undeformed geometry. In such cases we are able to express the force-displacement relationship for the discretized body in terms of Eq. (3.9). If, on the other hand, the displacements are large, then the deformed geometry will obviously differ significantly from the undeformed geometry. This results in a non-linear strain-displacement relationship. Large-displacement problems of this type are said to be 'geometrically non-linear'.

Geometrical non-linearity is a feature of 'elastic instability' problems which frequently occur in structural mechanics. In these problems, as the reader may be aware, a structure becomes unstable and eventually 'buckles' if the applied load exceeds its 'critical' or 'buckling' load. The Euler strut is a familiar example of this phenomenon. From the design point of view calculation of the 'critical' loads of structures is of considerable importance.

The finite element method can be used very effectively and in an elegant way to calculate critical loads[1,2,3]. As the strain-displacement relationship is now non-linear, the total strain within the deformed body can be expressed as the sum

$$\{\epsilon\}_{\text{total}} = \{\epsilon\} + \{\underline{\epsilon}\}^* \tag{14.1}$$

in which $\{\epsilon\}$ is the geometrically linear strain of the previous chapters. $\{\underline{\epsilon}\}$, on the other hand, is a non-linear function of displacement(s) and denotes the non-linear part of total strain.

Let H denote the strain energy of the deformed body due to $\{\epsilon\}$ and let $\underline{H}$ denote that due to $\{\underline{\epsilon}\}$. Then, the total strain energy of the body can be expressed as

$$H_{\text{total}} = H + \underline{H} \tag{14.2}$$

Throughout this chapter the material of the body will be assumed to obey Hooke's law. It is pointed out, however, that in practice both material and geometrical non-linearities may, and often do, occur simultaneously.

* Throughout this chapter underlined symbols will denote geometrically non-linear quantities.

14.2 Force-displacement relationship

By definition, H is given by

$$H = \frac{1}{2}\int_V \{\epsilon\}^{\mathrm{T}}\{\sigma\}\,\mathrm{d}v \tag{14.3}$$

Also, from section 9.8

$$\{\epsilon\} = [f]\{\bar{\delta}\}$$

and by definition

$$\{\sigma\} = [d]\{\epsilon\}$$

Consequently

$$H = \tfrac{1}{2}\{\bar{\delta}\}^{\mathrm{T}}\int_V [f]^{\mathrm{T}}[d][f]\,\mathrm{d}v\{\bar{\delta}\} \tag{14.4}$$

Then, applying the methods of matrix calculus, a differentiation of both sides of Eq. (14.4) with respect to $\{\bar{\delta}\}$ gives

$$\frac{\partial H}{\partial\{\bar{\delta}\}} = \int_V [f]^{\mathrm{T}}[d][f]\,\mathrm{d}v\{\bar{\delta}\} \tag{14.5}$$

[The rule for this differentiation is as follows:
If $\{x\}$ and $\{B\}$ are column vectors and $[A]$ a square matrix, and if

$$H = \tfrac{1}{2}\{x\}^{\mathrm{T}}[A]\{x\} + \{x\}^{\mathrm{T}}\{B\}$$

then,

$$\frac{\partial H}{\partial\{x\}} = [A]\{x\} + \{B\}$$]

From Eq. (9.11) the integral of this equation is recognized as matrix $[K_i]$. We can therefore write

$$\frac{\partial H}{\partial\{\bar{\delta}\}} = [K_i]\{\bar{\delta}\} \tag{14.6}$$

In order to derive a similar expression for $\underline{H}$, let $[\sigma_0]$ denote the *in-plane or axial* stresses within the element, as computed from the geometrically linear analysis. Also, let $\{\psi\}$ be such that the strain energy $\underline{H}$ can be expressed as

$$\underline{H} = \frac{1}{2}\int_V \{\psi\}^{\mathrm{T}}[\sigma_0]\{\psi\}\,\mathrm{d}v \tag{14.7}$$

[When, as in some cases, $[\sigma_0]$ and $\{\psi\}$ each denote a scalar quantity, Eq. (14.7) would become

$$\underline{H} = \frac{1}{2}\int_V \psi\sigma_0\psi\,\mathrm{d}v \tag{14.8}$$]

$\{\underline{\psi}\}$ is as yet unspecified, and will depend on the type of element under consideration as we shall soon see. Furthermore, let

$$\{\underline{\psi}\} = [\underline{N}]\{A\} = [\underline{N}][c^{-1}]\{\delta\} \tag{14.9}$$

in which $[\underline{N}]$ is the non-linear counterpart of $[N]$. Then, eliminating $\{\underline{\psi}\}$ between Eqs. (14.7) and (14.9), we have

$$\underline{H} = \tfrac{1}{2}\{\delta\}^{\mathrm{T}}[c^{-1}]^{\mathrm{T}}\int_V [\underline{N}]^{\mathrm{T}}[\sigma_0][\underline{N}]\,\mathrm{d}v[c^{-1}]\{\delta\} \tag{14.10}$$

A differentiation of both sides of Eq. (14.10) with respect to $\{\delta\}$ gives, as in Eq. (14.5),

$$\frac{\partial \underline{H}}{\partial\{\delta\}} = [c^{-1}]^{\mathrm{T}}\int_V [\underline{N}]^{\mathrm{T}}[\sigma_0][\underline{N}]\,\mathrm{d}v[c^{-1}]\{\delta\} \tag{14.11}$$

Introducing
$$[\underline{K}_i] = [c^{-1}]^{\mathrm{T}}\int_V [\underline{N}]^{\mathrm{T}}[\sigma_0][\underline{N}]\,\mathrm{d}v[c^{-1}] \tag{14.12}$$

Eq. (14.11) will now be written as

$$\frac{\partial \underline{H}}{\partial\{\delta\}} = [\underline{K}_i]\{\delta\} \tag{14.13}$$

From Eqs. (14.2), (14.6) and (14.13) it follows therefore that

$$\frac{\partial H_{\mathrm{total}}}{\partial\{\delta\}} = [K_i + \underline{K}_i]\{\delta\} \tag{14.14}$$

But, according to Castigliano's Theorem part I, the left hand side of Eq. (14.14) is equal to the forces acting on the body. Consequently, as in section 9.2, we can write this equation also as

$$\{\bar{P}\} = [K_i + \underline{K}_i]\{\delta\} \tag{14.15}$$

Matrix $[\underline{K}_i]$, which is derived by considering the strain energy due to $\{\underline{\epsilon}\}$, is called the 'geometrical stiffness matrix' of element *i*.

In the theory[4] of elastic instability $[\sigma_0]$ is assumed to be unaffected by transverse displacements; that is, it will remain constant with respect to $\{\underline{\epsilon}\}$, when $\{\underline{\epsilon}\}$ is a function of transverse displacement only. Consequently, if in a certain structure $[\sigma_0]$ is found to be equal to the single quantity σ_{xx}, for example, and if $\underline{\epsilon}$ is a function of transverse displacement only, then the strain energy per unit volume of the body due to $\underline{\epsilon}$ will clearly be equal to $\underline{\epsilon}\,.\,\sigma_{xx}$. This fact will be utilised for calculating $\underline{H}$ and subsequently $[\underline{K}_i]$.

Having calculated the $[K_i]$'s and $[\underline{K}_i]$'s of all elements into which the structure is subdivided, its buckling loads can be calculated as follows:

The $[K_i]$'s will be assembled in the usual way to form $[K]$; the $[\underline{K}_i]$'s will be

assembled in exactly the same way to form the 'overall geometrical stiffness matrix' $[\underline{K}]$ for the entire structure. The geometrically non-linear counterpart of Eq. (3.9) will then read

$$[K + \underline{K}]\{\delta\} = \{P\} \tag{14.16}$$

Now, introducing the scale factor λ, let us write

$$[\underline{K}] = \lambda[\underline{K}^*] \tag{14.17}$$

in which $[\underline{K}^*]$ is defined as the value of $[\underline{K}]$ when $\{P\}$ is equal to unity; Eq. (14.16) then becomes

$$[K + \lambda\underline{K}^*]\{\delta\} = \{P\} \tag{14.18}$$

and hence

$$\{\delta\} = [K + \lambda\underline{K}^*]^{-1}\{P\} \tag{14.19}$$

The 'buckling' loads of a structure are defined as applied loads at which its displacements $\{\delta\}$ tend to infinity. Eq. (14.19) shows that $\{\delta\}$ tends to infinity when its determinant vanishes, i.e., when

$$\det |[K] + \lambda[\underline{K}^*]| = 0 \tag{14.20}$$

Thus, if Eq. (14.20) has the roots $\lambda_1\ \lambda_2 \cdots \lambda_N$, then λ_k $(k = 1, 2 \cdots N)$ is the buckling load of the discretized structure in its k-th mode of buckling. Also, there is a buckling 'mode shape' associated with λ_k which represents in relative magnitudes the buckled shape of the structure in its k-th mode of buckling. From the design point of view the smallest root λ_1, which is the buckling load in the first or 'fundamental' mode of buckling, is of main interest.

Practical problems usually result in large $[K]$ and $[\underline{K}]$ matrices. It is a much more attractive proposition in such cases to determine the buckling loads, by using the computer, as the roots of the eigenvalue equation

$$[K]\{\delta\} = -\lambda[\underline{K}]\{\delta\} \tag{14.21}$$

which is an alternative statement of Eq. (14.20). The 'eigenvectors' of this equation are the buckling mode shapes.

14.3 The geometrical stiffness matrix of bar elements

Figure 14.1 shows the element under consideration. By virtue of assumed geometrical linearity, the transverse displacements of this element were negligible in section 9.7. In the present analysis, on the other hand, transverse displacement is significant* and therefore both axial and transverse displacements must be considered. Let

$$u(x) = A_0 + A_1 x \tag{14.22a}$$

and

$$v(x) = A_2 + A_3 x \tag{14.22b}$$

* Since the deformed geometry of the bar differs significantly from its undeformed geometry.

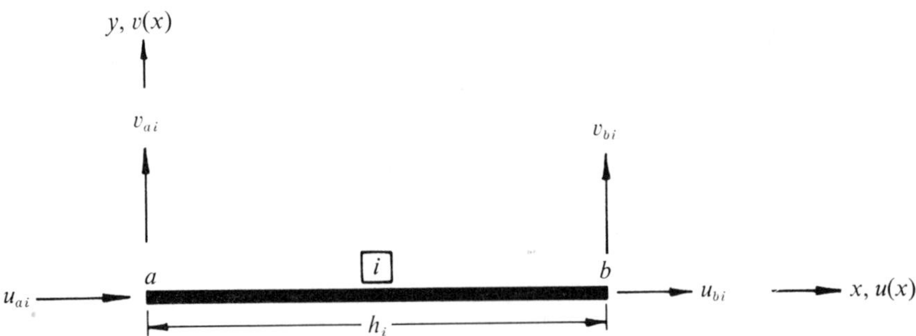

Fig. 14.1 A geometrically nonlinear uniform bar element.

be the assumed axial and transverse displacements respectively. Then, substituting the extremity coordinates into the above equations we have

$$u_{ai} = u(0) = A_0$$

$$v_{ai} = v(0) = A_2$$

$$u_{bi} = u(h_i) = A_0 + A_1 h_i$$

and

$$v_{bi} = v(h_i) = A_2 + A_3 h_i$$

As usual, we shall now put these equations in the matrix form of Eq. (9.23); thus, with

$$\{\delta\} = \begin{Bmatrix} u_{ai} \\ v_{ai} \\ u_{bi} \\ v_{bi} \end{Bmatrix} \quad \text{and} \quad \{A\} = \begin{Bmatrix} A_0 \\ A_1 \\ A_2 \\ A_3 \end{Bmatrix}$$

we can show that

$$[c] = \begin{bmatrix} 1 & 0 & 0 & 0 \\ 0 & 1 & 0 & 0 \\ 1 & h_i & 0 & 0 \\ 0 & 0 & 1 & h_i \end{bmatrix}$$

and, by inversion

$$[c^{-1}] = \begin{bmatrix} 1 & 0 & 0 & 0 \\ 0 & 1 & 0 & 0 \\ -1/h_i & 1/h_i & 0 & 0 \\ 0 & 0 & -1/h_i & 1/h_i \end{bmatrix} \tag{14.23}$$

It can be shown from non-linear elasticity theory that in the case of bars

$$\underline{\epsilon} = \frac{1}{2}\left(\frac{\partial v}{\partial x}\right)^2 \tag{14.24}$$

Now, let F_i be the axial force in the bar, as computed from the linear analysis. Then the stress within it is clearly

$$\sigma_{xx} = F_i/a_i \tag{14.25}$$

where a_i is the area of cross section.

Equation (14.24) shows that in this case $\underline{\epsilon}$ is a function of transverse displacement only. Consequently, as pointed out in section 14.2, σ_{xx} will remain constant with respect to $\underline{\epsilon}$. We can therefore write

$$\underline{H} = \int_V \epsilon\sigma_{xx}\,\mathrm{d}v = \frac{1}{2}\int_V (F_i/a_i)\left(\frac{\partial v}{\partial x}\right)^2 \mathrm{d}v \tag{14.26}$$

Then, writing this equation alternatively as

$$\underline{H} = \frac{1}{2}\int_V \left(\frac{\partial v}{\partial x}\right)(F_i/a_i)\left(\frac{\partial v}{\partial x}\right) \mathrm{d}v \tag{14.27}$$

and comparing Eq. (14.8) with (14.27), it is obvious that in this case

$$\psi = \frac{\partial v}{\partial x} \tag{14.28}$$

and

$$\sigma_0 = F_i/a_i \tag{14.29}$$

A differentiation of $v(x)$ of Eq. (14.22b) now gives

$$\psi = A_3 = \{0 \quad 0 \quad 0 \quad 1\}\{A\} \tag{14.30}$$

and, a comparison between Eqs. (14.9) and (14.30) shows that

$$\{\underline{N}\} = \{0 \quad 0 \quad 0 \quad 1\} \tag{14.31}$$

Then, from Eqs. (14.29) and (14.31) we can show that

$$\int_V [\underline{N}]^{\mathrm{T}}[\sigma_0][\underline{N}]\,\mathrm{d}v = F_i\int_0^{h_i}\begin{Bmatrix}0\\0\\0\\1\end{Bmatrix}\{0 \quad 0 \quad 0 \quad 1\}\,\mathrm{d}x = F_i h_i\begin{bmatrix}0&0&0&0\\0&0&0&0\\0&0&0&0\\0&0&0&1\end{bmatrix} \tag{14.32}$$

Finally, we can show from Eqs. (14.12), (14.23) and (14.32) that

$$[\underline{K}_i] = (F_i/h_i)\begin{bmatrix}0&0&0&0\\0&1&0&-1\\0&0&0&0\\0&-1&0&1\end{bmatrix} \tag{14.33}$$

Observe that since the size of $[\underline{K}_i]$ must be the same as that of $[K_i]$, the

latter (Eq. 7.2) will now be written as

$$[K_i] = k_i \begin{bmatrix} 1 & 0 & -1 & 0 \\ 0 & 0 & 0 & 0 \\ -1 & 0 & 1 & 0 \\ 0 & 0 & 0 & 0 \end{bmatrix} \tag{14.34}$$

14.4 Application to pin-jointed frames

In order to assemble the global version of $[\underline{K}]$ for the entire frame, the first step is obviously to calculate from the linear analysis the internal forces in the members of the frame. The $[\underline{K}_i]$'s can then be found from Eq. (14.33). The next step is to transform these so that they all refer to a common global system of coordinates. Here, as in Chapter 7, the global values of these matrices will be found from

$$[\underline{K}_i^g] = [T_i][\underline{K}_i][T_i]^T \tag{14.35}$$

In plane frames, for example, it is easily verified that the transformation matrix to be used in Eq. (14.35) is

$$[T_i] = \begin{bmatrix} l_i & -m_i & 0 & 0 \\ m_i & l_i & 0 & 0 \\ 0 & 0 & l_i & -m_i \\ 0 & 0 & m_i & l_i \end{bmatrix} \tag{14.36}$$

Consequently, we can show from Eqs. (14.33), (14.35) and (14.36) that

$$[\underline{K}_i^g] = (F_i/h_i) \begin{bmatrix} r & -s & -r & s \\ -s & q & s & -q \\ -r & s & r & -s \\ s & -q & -s & q \end{bmatrix} \tag{14.37}$$

where $q = l_i^2$, $r = m_i^2$ and $s = l_i m_i$ (l_i and m_i are defined in Fig. 7.1).

Having computed the $[\underline{K}_i^g]$'s, these will then be assembled in the usual way to form the 'global overall geometrical stiffness matrix' $[\underline{K}^g]$ of the whole frame. Then, with

$$[\underline{K}^g] = \lambda[\underline{K}^{g*}] \tag{14.38}$$

the buckling loads can be found as the roots of

$$\det |[K^g] + \lambda[\underline{K}^{g*}]| = 0 \tag{14.39}$$

in which $[K^g]$ is the linear global overall stiffness matrix of Chapter 7.

An example

Consider the simple plane frame of Fig. 14.2a, whose finite element model is shown in Fig. 14.2b. It is given that $a_1 = a_2 = a$ and $E_1 = E_2 = E$. Then,

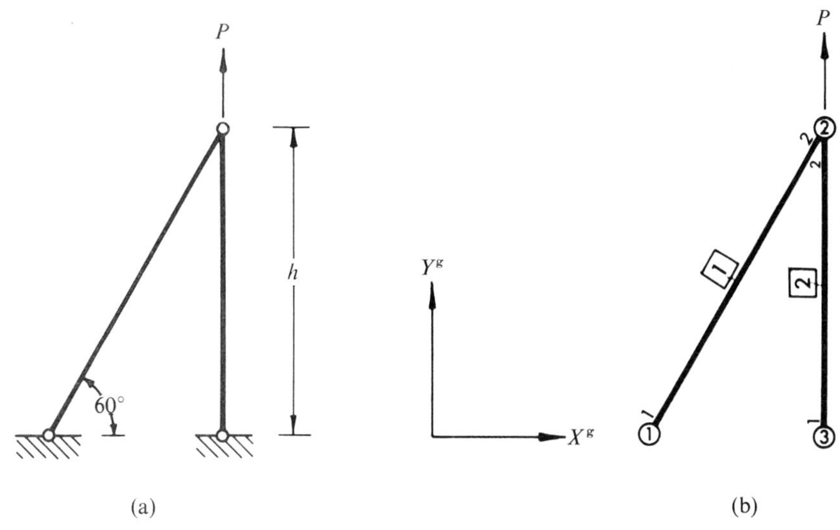

Fig. 14.2 (a) A pin-jointed plane frame. (b) Finite element model of (a).

from Eq. (7.12) we have

$$[K_1^g] = \frac{\sqrt{3}aE}{8h}\begin{bmatrix} 1 & \sqrt{3} & -1 & -\sqrt{3} \\ \sqrt{3} & 3 & -\sqrt{3} & -3 \\ -1 & -\sqrt{3} & 1 & \sqrt{3} \\ -\sqrt{3} & -3 & \sqrt{3} & 3 \end{bmatrix} \tag{14.40}$$

and

$$[K_2^g] = \frac{aE}{h}\begin{bmatrix} 0 & 0 & 0 & 0 \\ 0 & 1 & 0 & -1 \\ 0 & 0 & 0 & 0 \\ 0 & -1 & 0 & 1 \end{bmatrix} \tag{14.41}$$

The boundary conditions in this case are obviously

$$u_1^g = v_1^g = u_3^g = v_3^g = 0$$

Then, excluding nodes 1 and 3 and following the methods of Chapter 3 we can

show from Eqs. (14.40) and (14.41) that

$$[K^g] = \frac{\sqrt{3}aE}{8h}\begin{bmatrix} 1 & \sqrt{3} \\ \sqrt{3} & 3 + \dfrac{8}{\sqrt{3}} \end{bmatrix} \tag{14.42}$$

For calculating the $[\underline{K}_i^g]$'s from Eq. (14.37), we note that the internal member forces, as computed from the linear analysis, are $F_1 = 0$ and $F_2 = P$.* Consequently,

$$[\underline{K}_1^g] = [0]$$

and

$$[\underline{K}_2^g] = \frac{P}{h}\begin{bmatrix} 1 & 0 & -1 & 0 \\ 0 & 0 & 0 & 0 \\ -1 & 0 & 1 & 0 \\ 0 & 0 & 0 & 0 \end{bmatrix}$$

Assembling these matrices in the same way as the $[K_i^g]$'s, we now have

$$[\underline{K}^g] = \frac{P}{h}\begin{bmatrix} 1 & 0 \\ 0 & 0 \end{bmatrix}$$

or, noting that by definition $[\underline{K}^{g*}] = [\underline{K}^g]$ when $P = 1$, we obtain

$$[\underline{K}^{g*}] = \frac{1}{h}\begin{bmatrix} 1 & 0 \\ 0 & 0 \end{bmatrix} \tag{14.43}$$

Then, from Eqs. (14.39), (14.42) and (14.43)

$$\det\begin{vmatrix} 1 + \dfrac{8\lambda}{\sqrt{3}aE} & \sqrt{3} \\ \sqrt{3} & 3 + \dfrac{8}{\sqrt{3}} \end{vmatrix} = 0 \tag{14.44}$$

Solving this equation for λ, we obtain the buckling load as

$$P_{\text{critical}} = \lambda = -\frac{\sqrt{3}aE}{8 + 3\sqrt{3}}$$

The negative sign indicates that the buckling load acts in the direction opposite to that shown in Fig. 14.2a. This result agrees with the exact result due to Timoshenko and Gere[4]. The fact that Eq. (14.44) has a single root indicates that the frame has a single mode of buckling.

* F_i is positive when tensile and negative when compressive.

14.5 The geometrical stiffness matrix of beam elements

Consider the uniform beam element of Fig. 2.6 (axial and torsional modes are not relevant here, and will therefore be ignored). In this element, as in the bar element of section 14.4, we can show that

$$\sigma_0 = F_i/a_i$$

and

$$\underline{\epsilon} = \frac{1}{2}\left(\frac{\partial v}{\partial x}\right)^2$$

where $v(x)$ is the transverse deflection of the element. Consequently Eq. (14.28) is also valid for this element. Then, differentiating $v(x)$ of Eq. (9.32), we have

$$\begin{aligned}\psi &= A_1 + 2A_2x + 3A_3x^2 \\ &= \{0 \quad 1 \quad 2x \quad 3x^2\}\{A\}\end{aligned} \tag{14.45}$$

in which $\{A\}$ lists the constants of Eq. (9.32). Comparing Eqs. (14.9) and (14.45), it is clear that

$$\{\underline{N}\} = \{0 \quad 1 \quad 2x \quad 3x^2\} \tag{14.46}$$

Then, from Eqs. (14.29) and (14.46)

$$\begin{aligned}\int_V [\underline{N}]^{\mathrm{T}}[\sigma_0][\underline{N}]\,\mathrm{d}v &= F_i\int_0^{h_i}\begin{Bmatrix}0\\1\\2x\\3x^2\end{Bmatrix}\{0 \quad 1 \quad 2x \quad 3x^2\}\,\mathrm{d}x \\ &= F_i\int_0^{h_i}\begin{bmatrix}0 & 0 & 0 & 0\\ 0 & 1 & 2x & 3x^2\\ 0 & 2x & 4x^2 & 6x^3\\ 0 & 3x^2 & 6x^3 & 9x^4\end{bmatrix}\mathrm{d}x \\ &= F_ih\begin{bmatrix}0 & 0 & 0 & 0\\ 0 & 1 & h & h^2\\ 0 & h & \dfrac{4h^2}{3} & \dfrac{3h^3}{2}\\ 0 & h^2 & \dfrac{3h^3}{2} & \dfrac{9h^4}{5}\end{bmatrix}\end{aligned} \tag{14.47}$$

in which we have written h for h_i for convenience. Using Eqs. (9.35) and (14.47)

in Eq. (14.12), it can be verified by multiplication that

$$[\underline{K}_i] = (F_i/h)\begin{bmatrix} \frac{6}{5} & & \text{symm.} & \\ \frac{h}{10} & \frac{2h^2}{15} & & \\ -\frac{6}{5} & -\frac{h}{10} & \frac{6}{5} & \\ \frac{h}{10} & -\frac{h^2}{30} & -\frac{h}{10} & \frac{2h^2}{15} \end{bmatrix} \tag{14.48}$$

When the $[K_i]$ of the element contains axial and/or torsional stiffnesses (e.g., Eq. 2.8), zeros must be introduced at the appropriate locations in $[\underline{K}_i]$, so that both these matrices are of the same size. Also, when, as in rigid-jointed frames, the elements are differently aligned, $[\underline{K}_i]$ of Eq. (14.48) will have to be transformed by using the transformation matrices of Chapter 6.

An example

As a typical application of $[\underline{K}_i]$ of Eq. (14.48), consider the axially loaded uniform strut of Fig. 14.3a. To begin with, let us represent this strut by the single element of Fig. 14.3b. Then, we can show from Eq. (2.8) (ignoring axial and torsional modes) that since transverse displacement is zero at both ends of the strut

$$[K] = \frac{EI}{h}\begin{bmatrix} 4 & 2 \\ 2 & 4 \end{bmatrix} \tag{14.49}$$

Similarly, from Eq. (14.48) we obtain

$$[\underline{K}] = -Ph\begin{bmatrix} \frac{2}{15} & -\frac{1}{30} \\ -\frac{1}{30} & \frac{2}{15} \end{bmatrix} \tag{14.50}$$

and hence by definition, with $P = 1$, we have

$$[\underline{K}^*] = -h\begin{bmatrix} \frac{2}{15} & -\frac{1}{30} \\ -\frac{1}{30} & \frac{2}{15} \end{bmatrix} \tag{14.51}$$

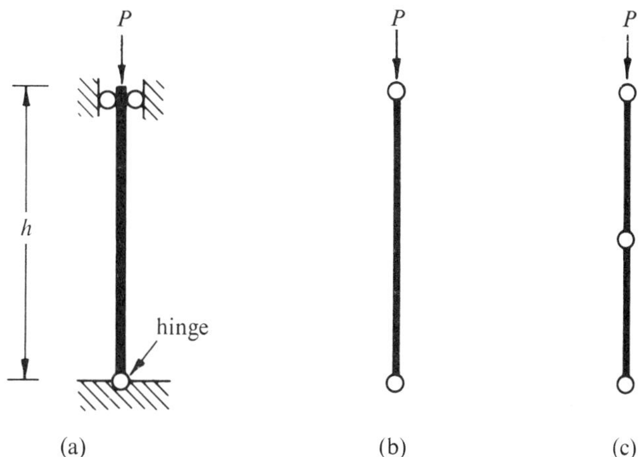

Fig. 14.3 (a) A uniform strut (EI constant). (b) Single element representation of (a). (c) Two-element representation of (a).

Consequently, Eq. (14.20) becomes

$$\det \begin{vmatrix} 4\beta - \dfrac{2h\lambda}{15} & 2\beta + \dfrac{h\lambda}{30} \\ 2\beta + \dfrac{h\lambda}{30} & 4\beta - \dfrac{2h\lambda}{15} \end{vmatrix} = 0$$

where $\beta = EI/h$. The roots of this equation are: $\lambda_1 = 12EI/h^2$ and $\lambda_2 = 60EI/h^2$. The buckling load in the fundamental buckling mode is therefore

$$P_{\text{critical}} = \lambda_1 = 12EI/h^2$$

The exact value of this load is $\pi^2 EI/h^2$; our result is thus in error by 21·6%. This is to be expected, however, since the strut was represented by a single finite element. Accuracy will be found to improve when the strut is represented by multiple elements; it can be verified, for instance, that when it is represented by two elements, as in Fig. 14.3c, the error is a mere 2·7%.

The larger root λ_2 is equal to the value of P_{critical} in the second buckling mode of the strut. The computed value of this load here is over 1·5 times its exact value of $4\pi^2 EI/h^2$. However, this value, like that of λ_1, would improve as we increase the number of elements representing the strut.

14.6 Application to plates

In the large-displacement theory of thin flat plates it is shown[5] that the non-linear strain $\{\underline{\epsilon}\}$ within a deformed plate, whose 'middle plane' (Fig. 10.1)

coincides with the $x - y$ plane in the undeformed state, is

$$\{\underline{\epsilon}\} = \begin{Bmatrix} \underline{e}_{xx} \\ \underline{e}_{yy} \\ \underline{e}_{xy} \end{Bmatrix} = \begin{Bmatrix} \frac{1}{2}\left(\frac{\partial w}{\partial x}\right)^2 \\ \frac{1}{2}\left(\frac{\partial w}{\partial y}\right)^2 \\ \frac{\partial w}{\partial x}\cdot\frac{\partial w}{\partial y} \end{Bmatrix} \tag{14.52}$$

where $w(x, y)$ denotes transverse deflection of the plate. It follows therefore that if the stresses within the element, as computed from the geometrically linear analysis, are

$$\sigma'_{xx},\ \sigma'_{yy}\ \text{ and }\ \tau'_{xy},$$

then,

$$\underline{H} = \int_V \left(\tfrac{1}{2}\sigma'_{xx}\left(\frac{\partial w}{\partial x}\right)^2 + \tfrac{1}{2}\sigma'_{yy}\left(\frac{\partial w}{\partial y}\right)^2 + \tau'_{xy}\left(\frac{\partial w}{\partial x}\right)\left(\frac{\partial w}{\partial y}\right)\right) \mathrm{d}v$$

This equation can be written in the matrix form as

$$\underline{H} = \frac{1}{2}\int_V \begin{Bmatrix} \frac{\partial w}{\partial x} \\ \frac{\partial w}{\partial y} \end{Bmatrix}^{\mathrm{T}} \begin{bmatrix} \sigma'_{xx} & \tau'_{xy} \\ \tau'_{xy} & \sigma'_{yy} \end{bmatrix} \begin{Bmatrix} \frac{\partial w}{\partial x} \\ \frac{\partial w}{\partial y} \end{Bmatrix} \mathrm{d}v \tag{14.53}$$

Comparing Eq. (14.7) with Eq. (14.53) it is clear that in this case

$$[\sigma_0] = \begin{bmatrix} \sigma'_{xx} & \tau'_{xy} \\ \tau'_{xy} & \sigma'_{yy} \end{bmatrix} \tag{14.54}$$

and

$$\{\psi\} = \begin{Bmatrix} \frac{\partial w}{\partial x} \\ \frac{\partial w}{\partial y} \end{Bmatrix} \tag{14.55}$$

The $[\underline{N}]$ matrix of the element can now be found from Eq. (14.55); from Eqs. (10.21), (14.9) and (14.55) we can demonstrate that $[\underline{N}]$ for the triangular plate element, for example, is

$$[\underline{N}] = \begin{bmatrix} 0 & 1 & 0 & 2x & y & 0 & 3x^2 & (2xy + y^2) & 0 \\ 0 & 0 & 1 & 0 & x & 2y & 0 & (2xy + x^2) & 3y^2 \end{bmatrix} \tag{14.56}$$

Noting that in the case of a thin plate element of uniform thickness t Eq. (14.12) becomes

$$[\underline{K}_i] = t \,.\, [c^{-1}]^{\mathrm{T}} \int_S [\underline{N}]^{\mathrm{T}}[\sigma_0][\underline{N}]\,\mathrm{d}x\,\mathrm{d}y[c^{-1}]$$

in which the integration is performed over the entire area of the element, the $[\underline{K}_i]$ matrix of the triangular plate element can now be found by substituting $[\sigma_0]$, $[N]$ and $[c^{-1}]$ (obtained by inverting $[c]$ of Table 10.3) into the above equation. The procedure for the rectangular plate element is identical.

Equations (11.18) and (11.19) show that the strains and consequently $[\sigma_0]$ within a rectangular element varies in a linear fashion. In such cases, it is necessary therefore to evaluate the closed form of the integral

$$\int_S [\underline{N}]^{\mathrm{T}}[\sigma_0][\underline{N}]\,\mathrm{d}x\,\mathrm{d}y$$

for every element of the discretized plate, having taken the variation of $[\sigma_0]$ into consideration. However, this being a rather tedious process, it is often convenient, without significant loss of accuracy, to set in each element σ'_{xx}, σ'_{yy} and τ'_{xy} equal to their respective values averaged over that element.

14.7 Transverse vibration of structures subjected to axial or in-plane forces

It is well known that when subjected to axial or in-plane forces, the dynamic characteristics of plates, beams etc. are modified—the extent of modification depending upon the magnitude and sense of these forces. The finite element method can be used to determine accurately the natural frequencies and mode shapes of these structures.

It was demonstrated in section 13.1 that the natural frequencies of a geometrically linear structure were the roots of Eq. (13.1.22). If the aspect of geometrical non-linearity is introduced, then it is clear that the total stiffness of the structure becomes the sum

$$[K] + [\underline{K}]$$

As a result, the geometrically non-linear counterpart of Eq. (13.1.22) now reads

$$\det |[K + \underline{K}] - \omega^2[m]| = 0 \tag{14.57}$$

or, in terms of an eigenvalue equation

$$[K + \underline{K}]\{\delta\} = \omega^2[m]\{\delta\} \tag{14.58}$$

These equations can now be solved to calculate the natural frequencies and mode shapes of the structure under consideration.

An example

As a typical application of the above equations, let us calculate the fundamental frequency of transverse vibration of the axially loaded strut of Fig. 14.3a. Representing the strut by a single element, as in Fig. 14.3b, we have from Eqs. (14.49) and (14.50)

$$[K + \underline{K}] = \begin{bmatrix} 4\beta - \dfrac{2Ph}{15} & 2\beta + \dfrac{Ph}{30} \\ 2\beta + \dfrac{Ph}{30} & 4\beta - \dfrac{2Ph}{15} \end{bmatrix} \tag{14.59}$$

As for the $[m]$ matrix of the strut, we note that in the single element approximation it is equal to $[m_i]$ of Eq. (13.1.16). Furthermore, since transverse deflection is zero at both ends of the strut, rows 1 and 3 of $[m]$ and the corresponding columns will now be deleted. Consequently,

$$[m] = \frac{\rho a h^3}{420}\begin{bmatrix} 4 & -3 \\ -3 & 4 \end{bmatrix} \tag{14.60}$$

where a denotes area of cross section. Then, from Eqs. (14.57), (14.59) and (14.60) the frequency equation of this strut is found to be

$$\det \begin{vmatrix} 4\beta - \dfrac{2Ph}{15} - \dfrac{\gamma}{105} & 2\beta + \dfrac{Ph}{30} + \dfrac{\gamma}{140} \\ 2\beta + \dfrac{Ph}{30} + \dfrac{\gamma}{140} & 4\beta - \dfrac{2Ph}{15} - \dfrac{\gamma}{105} \end{vmatrix} = 0 \tag{14.61}$$

where $\gamma = \omega^2 \rho a h^3$. Expanding the above determinant and solving, we can show that the two roots of this equation give

$$\omega_1 = 10{\cdot}954\left(\frac{EI}{\rho a h^4}\right)^{\frac{1}{2}}\left(1 - \frac{Ph^2}{12EI}\right)^{\frac{1}{2}} \tag{14.62}$$

and

$$\omega_2 = 50{\cdot}200\left(\frac{EI}{\rho a h^4}\right)^{\frac{1}{2}}\left(1 - \frac{Ph^2}{60EI}\right)^{\frac{1}{2}} \tag{14.63}$$

which are the fundamental and the first harmonic (natural) circular frequencies respectively. The exact values of these frequencies are:

$$\omega_1 = \pi^2\left(\frac{EI}{\rho a h^4}\right)^{\frac{1}{2}}\left(1 - \frac{Ph^2}{\pi^2 EI}\right)^{\frac{1}{2}}$$

and

$$\omega_2 = 4\pi^2\left(\frac{EI}{\rho a h^4}\right)^{\frac{1}{2}}\left(1 - \frac{Ph^2}{4\pi^2 EI}\right)^{\frac{1}{2}}$$

These values of ω_1 and ω_2 show that:

(a) The computed results are obviously in substantial error. However, in general, they will improve and approach their exact values as the strut is subdivided into larger number of elements.

(b) Natural frequencies diminish when the axial force is compressive, as in Fig. 14.3a, and increase when it is tensile.

(c) Equation (14.62) shows that $\omega_1 = 0$ when

$$P = 12EI/h^2$$

But this value of P is the computed value of P_{critical} for the fundamental buckling mode of the strut (see section 14.5). Similarly, $\omega_2 = 0$, as Eq. (14.63) shows, when P becomes equal to the computed value of P_{critical} for the strut's second buckling mode. We can therefore make the following statement:

The structure ceases to vibrate in its k-th vibration mode (i.e., $\omega_k = 0$; see section 13.1.3), if the applied axial or in-plane force is compressive and its magnitude is equal to that of P_{critical} for the k-th buckling mode; this will happen provided that the structure's k-th vibration and buckling modes are identical. When, as in some cases, they are not identical, the situation becomes much more complex.

(a) and (b) apply to all and (c) to most structures of this type.

References

1. Argyris, J. H., Kelsey, S. and Kamel, H., 'Matrix Methods in Structural analysis, AGARD-ograph 72, Pergamon Press, pp. 1–164, 1964.

2. Turner, M. J., Martin, H. C. and Weikel, R. C., 'Further Development and Application of the Stiffness method', AGARD-ograph 72, Pergamon Press, pp. 203–66, 1964.

3. Martin, H. C., 'On the derivation of stiffness matrices for the analysis of large deflection and stability problems', *Proc. conf. Matrix Methods in Structural Mechanics*, Air Force Base Institute of Technology, Wright Patterson Air Force Base, Ohio, Oct. 1965.

4. Timoshenko, S. P. and Gere, J. M., *Theory of Elastic Stability*, 2nd edn., McGraw-Hill, 1961.

5. Fung, Y. C., *Foundations of Solid Mechanics*, Prentice Hall Inc., 1965.

15 Computation

15.1 General

Our primary objective in this chapter is to introduce the reader, who is assumed to possess elementary knowledge of FORTRAN programming, to the fundamentals of programming for finite elements. As we have already remarked, scientific and engineering problems, when discretized by finite elements usually lead to large overall stiffness matrices whose processing makes the use of a computer inevitable. Owing to its high speed and general versatility, the modern digital computer is ideally suited for the numerical solution of most scientific and engineering problems, provided that we are able to supply it with all relevant data and the scheme of solution.

Computer solution of finite element problems is considerably facilitated when all essential data are organized in a systematic way. Perhaps the most important data in this regard are those concerned with the assembly of the overall stiffness matrix; these will be given particular emphasis here, since, as experience shows, nearly all beginners find this process most difficult to understand and program for. This, as well as all other computational steps leading up to the solution of the problem will be presented in simple terms by devising separate sets of data required for each individual operation.

The tendency to write general purpose programs—the so called 'package' programs—will be deliberately discouraged, as these programs almost certainly inhibit the beginner from understanding the vital processes involved in programming for finite elements. It is also pointed out that the particular manner of programming and data organization given here is by no means unique; on the contrary, there are other possible ways which one could adopt, once the fundamentals have been assimilated.

However, before embarking on the aspects of computation, we shall discuss in the next two sections some important mathematical topics which are relevant in the context of finite element analysis.

15.2 Methods for improving $[K_i]$

There is a number of ways in which the $[K_i]$ matrix of a finite element can be improved for better accuracy; of these the following are widely used:

The method of 'fictitious' nodes

One or more 'fictitious' nodes, defined as internal nodes at which no external force is applied, are used in this method. Consider the triangular element *nkj*

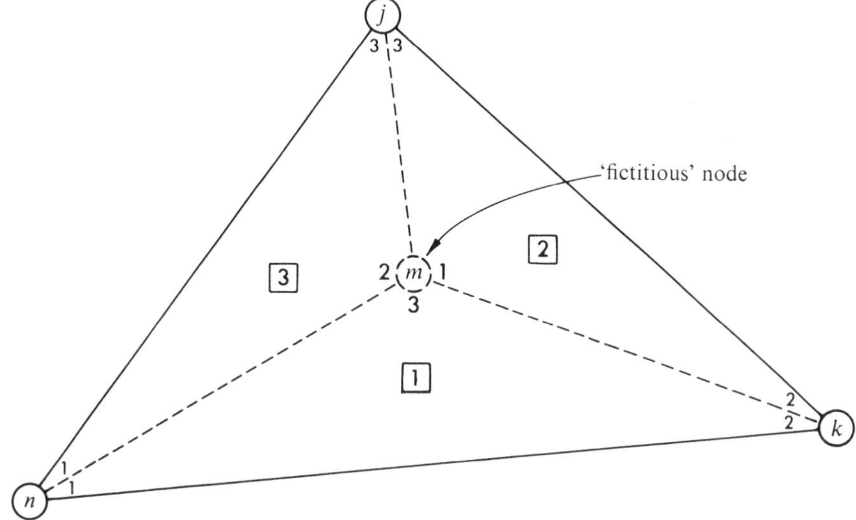

Fig. 15.1 A triangular element with a 'fictitious' node.

(Fig. 15.1) containing the single fictitious node m, which subdivides the element into the smaller triangles 1, 2 and 3 (note that circled and squared digits denote node and element numbers respectively). These triangles and the nodes connecting them can now be imagined to constitute a complete 'structure', whose overall stiffness matrix we can assemble by the methods of Chapter 3. Indeed, for this 'structure' we can show that Eq. (3.9) becomes

$$\begin{Bmatrix} \{P_n\} \\ \{P_k\} \\ \{P_j\} \\ \hline \{P_m\} \end{Bmatrix} = \left[\begin{array}{ccc:c} S(1,1,1) + S(3,1,1) & S(1,1,2) & S(3,1,3) & S(3,1,2) + S(1,1,3) \\ S(1,2,1) & S(1,2,2) + S(2,2,2) & S(2,2,3) & S(1,2,3) + S(2,2,1) \\ S(3,3,1) & S(2,3,2) & S(2,3,3) + S(3,3,3) & S(2,3,1) + S(3,3,2) \\ \hdashline S(3,2,1) + S(1,3,1) & S(1,3,2) + S(2,1,2) & S(2,1,3) + S(3,2,3) & S(1,3,3) + S(2,1,1) + S(3,2,2) \end{array}\right] \begin{Bmatrix} \{\delta_n\} \\ \{\delta_k\} \\ \{\delta_j\} \\ \hline \{\delta_m\} \end{Bmatrix} \tag{15.1}$$

in which the $\{P_k\}$'s and the $\{\delta_k\}$'s are defined in section 2.7. It is understood here that the $S(i, J, K)$'s of elements 1, 2 and 3 are explicitly known. Let us now write this equation in the symbolic form as

$$\begin{Bmatrix} \{P\} \\ \hline \{P_m\} \end{Bmatrix} = \left[\begin{array}{c:c} A & B \\ \hdashline B^{\mathrm{T}} & C \end{array}\right] \begin{Bmatrix} \{\delta\} \\ \hline \{\delta_m\} \end{Bmatrix} \tag{15.2}$$

The equivalence between the matrices of Eq. (15.1) and those of Eq. (15.2) is obvious. Now, since node m is fictitious, we have by definition $\{P_m\} = \{0\}$.

Equation (15.2) can therefore be multiplied out to give

$$\{P\} = [A]\{\delta\} + [B]\{\delta_m\} \tag{15.3a}$$

and

$$\{0\} = [B]^{\mathrm{T}}\{\delta\} + [C]\{\delta_m\} \tag{15.3b}$$

Premultiplying both sides of Eq. (15.3b) by $[C^{-1}]$ and then eliminating $\{\delta_m\}$ between the result and Eq. (15.3a), we can show that

$$\{P\} = [A - BC^{-1}B^{\mathrm{T}}]\{\delta\} \tag{15.4}$$

Clearly, Eq. (15.4) relates the external forces at the nodes n, k and j to the displacements at these nodes. Moreover, as we have considered the element in isolation, its nodal forces and displacements are respectively equal to its extremity (internal) forces and displacements. Consequently, by definition, we can write

$$[K_i]_{\text{improved}} = [A - BC^{-1}B^{\mathrm{T}}] \tag{15.5}$$

Whereas $[K_i]$ is derived from the characteristics of the triangle nkj, an inspection of Fig. 15.1 and Eq. (15.5) shows that the derivation of $[K_i]_{\text{improved}}$ is based on a finer subdivision of nkj into the smaller elements 1, 2 and 3. Therefore, according to the remarks made in section 9.6, results due to $[K_i]_{\text{improved}}$ will be more accurate than those due to $[K_i]$. This procedure can obviously be applied to all finite elements.

The above process is essentially one in which interior nodes, which are free from external forces, are eliminated. We can therefore also employ this method to reduce the size of a given problem by eliminating from it all such nodes. This will obviously lead to considerable saving in computer storage requirement. Consider, for instance, the nodes of Fig. 15.2 which form one of the 'sub-regions' into which a large problem can be divided. Let the degree of freedom be 3 and let the nodes 6, 7 and 8 be free from external forces. Now, if we assemble the $[K]$ matrix for this sub-region in the usual way, then its size will be 24×24. If, on the other hand, nodes 6, 7 and 8 are treated as 'fictitious' and eliminated by the method just described, then $[K]$ will have the much reduced size of 15×15. Clearly, if this process were repeated for all the sub-regions of the problem, then this would result in a considerable saving in computer storage. The computer can easily be programmed to carry out this type of node elimination entirely automatically.

The method of additional extremities

Another commonly used method for improving $[K_i]$ is by introducing a number of 'additional extremities' into the element. Consider, for example, the triangular element of Fig. 15.3a (ignore the dotted lines for the time being) in which d, e and f are the additional extremities. Therefore, according to section 9.5(a), $n = 6$ in this case. As a result, we can assume each mutually independent

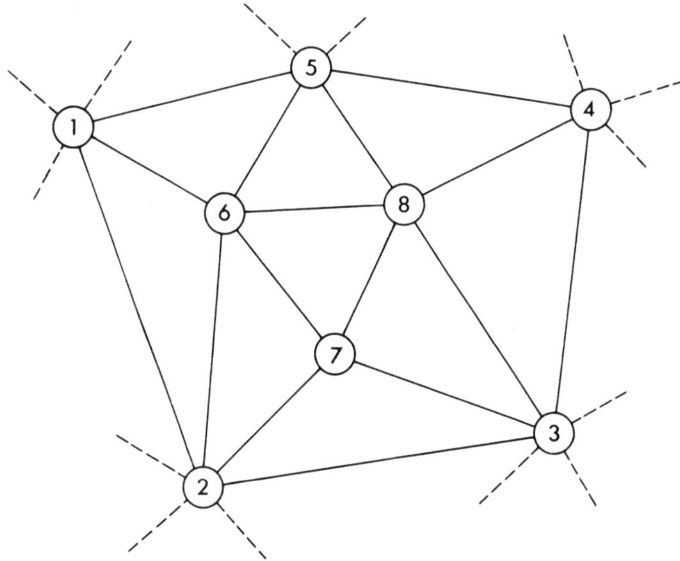

Fig. 15.2 A typical 'sub-region' of a large problem.

displacement component of the element as a polynomial defined by 6 unknown constants (recall that in the usual 3-extremity triangular element 3 constants could be used for each component).

As an example, suppose that the element of Fig. 15.3a is for the analysis of plane problems (Chapter 11). Then, using 6 constants to define each component

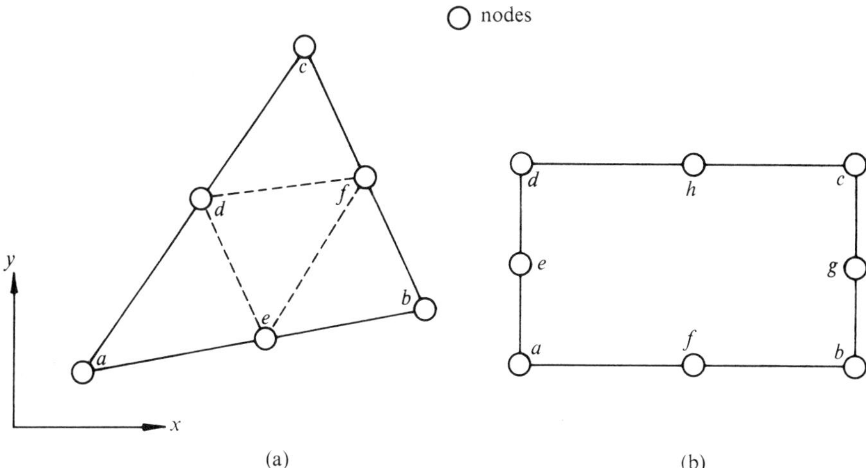

Fig. 15.3 Typical 'multi-extremity' elements; (a) triangular, *d*, *e* and *f* are additional extremities, (b) rectangular, *e*, *f*, *g* and *h* are additional extremities.

of displacement, we may assume

$$u(x, y) = A_0 + A_1x + A_2y + A_3xy + A_4x^2 + A_5y^2 \tag{15.6a}$$

and

$$v(x, y) = A_6 + A_7x + A_8y + A_9xy + A_{10}x^2 + A_{11}y^2 \tag{15.6b}$$

As usual, the coordinates (x_j, y_j) of extremity j $(= a, b, c, \ldots, f)$ are now substituted into these equations to obtain the extremity displacements u_{ji} and v_{ji}. All these displacements are then arranged in the matrix form of Eq. (11.2) to give $[c]$, which will emerge in this case as a 12×12 matrix. The $[N]$ matrix can be found, as usual, by differentiating Eqs. (15.6), in this case according to the scheme of Eq. (11.4a). With $[c]$ and $[N]$ thus found, we can now proceed to calculate $[K_i]$ of the element from the equations of section 9.8.

An inspection shows that unlike the linear variation of displacements represented by Eqs. (10.30), Eqs. (15.6) represent a parabolic variation which obviously approximates the true displacements of the element relatively more closely. Consequently, $[K_i]$, derived from Eqs. (15.6), is capable of more accurate results than when it is derived from Eqs. (10.30). It is easy to see as to why this should be so: referring to Fig. 15.3a, we find that the introduction of the additional extremities may be viewed as amounting to a finer subdivision of the original element into a number of smaller elements, as indicated by the dotted lines; hence the improved accuracy. In fact, experience shows that usually slightly better results are obtained by using additional extremities, rather than a finer subdivision which results when these extremities are joined together as shown by the dotted lines in Fig. 15.3a.

This method can obviously be used in any finite element; a typical multi-extremity rectangular element is shown in Fig. 15.3b. Furthermore, the matrices $[\underline{K}_i]$ and $[m_i]$ can also be improved for better accuracy by using this method.

15.3 Indirect inversion of a matrix

In the earlier chapters we have often seen that the $[c]$ matrix could not be directly inverted, as it had one or more zeros on the leading diagonal. However, $[c]$ or any other square matrix, whose determinant does not vanish, can be indirectly inverted by the following methods which are easily programmed for the computer:

The method of 'row interchange'

Consider the set of equations

$$\begin{aligned} 4b_1 + 2b_2 - b_3 + b_4 &= 9 \\ 2b_1 + 3b_3 - b_4 &= 7 \\ -b_1 + b_2 + b_3 + 2b_4 &= 12 \\ 3b_1 - b_2 + 2b_3 + 5b_4 &= 27 \end{aligned} \tag{15.7}$$

which can be put in the matrix form as

$$[A]\begin{Bmatrix} b_1 \\ b_2 \\ b_3 \\ b_4 \end{Bmatrix} = \begin{Bmatrix} 9 \\ 7 \\ 12 \\ 27 \end{Bmatrix} \tag{15.8}$$

where

$$[A] = \begin{array}{c} 1 \\ 2 \\ 3 \\ 4 \\ \\ \end{array}\!\!\begin{array}{c} \left[\begin{array}{rrrr} 4 & 2 & -1 & 1 \\ 2 & 0 & 3 & -1 \\ -1 & 1 & 1 & 2 \\ 3 & -1 & 2 & 5 \end{array}\right] \\ \begin{array}{rrrr} 1 & 2 & 3 & 4 \end{array} \end{array} \tag{15.9}$$

Since the determinant of $[A]$ is not equal to zero, the solution of Eq. (15.8) is obviously

$$\begin{Bmatrix} b_1 \\ b_2 \\ b_3 \\ b_4 \end{Bmatrix} = [A^{-1}]\begin{Bmatrix} 9 \\ 7 \\ 12 \\ 27 \end{Bmatrix}$$

We are unable to invert $[A]$ directly, however, as it has a zero on its leading diagonal in row 2. Let us therefore 'recondition' $[A]$ by interchanging the 'offending' row with another row, so that no zero appears in the leading diagonal of reconditioned $[A]$. Interchanging rows 2 and 3, for example, we can write Eqs. (15.7) in the matrix form as

$$[A_r]\begin{Bmatrix} b_1 \\ b_2 \\ b_3 \\ b_4 \end{Bmatrix} = \begin{Bmatrix} 9 \\ 12 \\ 7 \\ 27 \end{Bmatrix}$$

in which $[A_r]$ is the 'reconditioned' version of $[A]$, and is given by

$$[A_r] = \begin{array}{c} 1 \\ 3 \\ 2 \\ 4 \\ \\ \end{array}\!\!\begin{array}{c} \left[\begin{array}{rrrr} 4 & 2 & -1 & 1 \\ -1 & 1 & 1 & 2 \\ 2 & 0 & 3 & -1 \\ 3 & -1 & 2 & 5 \end{array}\right] \\ \begin{array}{rrrr} 1 & 2 & 3 & 4 \end{array} \end{array}$$

We are now able to invert $[A_r]$ directly, and it can verified that

$$[A_r^{-1}] = \begin{bmatrix} 0{\cdot}12500 & -0{\cdot}18750 & 0{\cdot}06250 & 0{\cdot}06250 \\ 0{\cdot}20833 & 0{\cdot}39583 & 0{\cdot}06250 & -0{\cdot}18750 \\ -0{\cdot}08333 & 0{\cdot}16667 & 0{\cdot}25000 & 0{\cdot}00000 \\ 0{\cdot}00000 & 0{\cdot}12500 & -0{\cdot}12500 & 0{\cdot}12500 \end{bmatrix}$$
$$\begin{matrix} \quad 1 & \quad\quad 2 & \quad\quad 3 & \quad\quad 4 \end{matrix}$$

Matrix $[A^{-1}]$ will now be formed simply by interchanging the *columns* 2 and 3 of $[A_r^{-1}]$. Thus,

$$[A^{-1}] = \begin{bmatrix} 0{\cdot}12500 & 0{\cdot}06250 & -0{\cdot}18750 & 0{\cdot}06250 \\ 0{\cdot}20833 & 0{\cdot}06250 & 0{\cdot}39583 & -0{\cdot}18750 \\ -0{\cdot}08333 & 0{\cdot}25000 & 0{\cdot}16667 & 0{\cdot}00000 \\ 0{\cdot}00000 & -0{\cdot}12500 & 0{\cdot}12500 & 0{\cdot}12500 \end{bmatrix} \tag{15.10}$$
$$\begin{matrix} \quad 1 & \quad\quad 3 & \quad\quad 2 & \quad\quad 4 \end{matrix}$$

Obviously, this method can also be used when the leading diagonal contains multiple zeros, provided of course that we are able to recondition the matrix in the manner indicated above by interchanging rows.

When it is convenient to do so, the columns of $[A]$ instead of its rows will be interchanged in a similar manner to form $[A]^{-1}$. Also, sometimes it may be expedient to adopt a 'mixed' scheme in which both rows and columns of $[A]$ are interchanged simultaneously.

The $[Z]$ *matrix method*

Alternatively, $[A]$ of Eq. (15.9) can be inverted as follows:

Let $[Z]$ be a square matrix whose determinant is not equal to zero, and let

$$[X] = [A][Z] \tag{15.11}$$

Then, $[X]$ can be directly inverted if the non-zero elements of $[Z]$ are chosen in such a way that no zero appears on the leading diagonal of $[X]$. An inversion of both sides of Eq. (15.11) now gives since $[[A][Z]]^{-1} = [Z^{-1}][A^{-1}]$

$$[X^{-1}] = [Z^{-1}][A^{-1}] \tag{15.12}$$

and a premultiplication of both sides of Eq. (15.12) by $[Z]$ leads to

$$[A^{-1}] = [Z][X^{-1}] \tag{15.13}$$

Equation (15.13) shows that $[A^{-1}]$ will be formed by first inverting $[X]$, defined by Eq. (15.11), and then by forming the product $[Z][X^{-1}]$.

As a typical application of this method, consider again the inversion of $[A]$

(Eq. 15.9). If we choose $[Z]$ arbitrarily as

$$[Z] = \begin{bmatrix} 1 & 1 & 0 & 0 \\ 0 & 1 & 0 & 0 \\ 0 & 0 & 1 & 0 \\ 0 & 0 & 0 & 1 \end{bmatrix}$$

then from Eqs. (15.9) and (15.11), we have

$$[X] = [A][Z] = \begin{bmatrix} 4 & 6 & -1 & 1 \\ 2 & 2 & 3 & -1 \\ -1 & 0 & 1 & 2 \\ 3 & 2 & 2 & 5 \end{bmatrix}$$

Since no element on the leading diagonal of $[X]$ is zero, we can directly invert it to give

$$[X^{-1}] = \begin{bmatrix} -0{\cdot}08333 & 0{\cdot}00000 & -0{\cdot}58333 & 0{\cdot}25000 \\ 0{\cdot}20833 & 0{\cdot}06250 & 0{\cdot}39583 & -0{\cdot}18750 \\ -0{\cdot}08333 & 0{\cdot}25000 & 0{\cdot}16667 & 0{\cdot}00000 \\ 0{\cdot}00000 & -0{\cdot}12500 & 0{\cdot}12500 & 0{\cdot}12500 \end{bmatrix}$$

It can now be verified by multiplication that the product $[Z][X^{-1}]$ is exactly equal to $[A^{-1}]$ of Eq. (15.10). Observe that the choice of $[Z]$ is arbitrary, the only requirement placed upon it being that the product $[A][Z]$ must not contain any zero on its leading diagonal. Experience shows that best results are obtained by setting

$$[Z] = [A]^{\mathrm{T}}$$

This method, like that of row interchange, is a general one in that it can be used to invert any matrix which cannot be directly inverted and whose determinant does not vanish.

15.4 Organization of input data

The computer processing of a typical finite element problem normally consists of four distinct operations, which are executed in the following order:

(1) Calculation of $[K_i]$ for every element representing the body,
(2) Assembly of the overall stiffness matrix $[K]$ and definition of the overall external force vector $\{P\}$,
(3) Solution of an equation similar to Eq. (3.9) for $\{\delta\}$, and
(4) Calculation of stresses and strains etc.

Sufficient 'input' data must be provided for the computer to carry out operations 1, 2 and 4 while, operation 3 requires a device for solving a set of linear simultaneous equations, the general matrix form of which is represented by Eq. (3.9). Such a device can be found in a large variety of 'subroutines' normally available in all program libraries. The particular subroutine to be used in a given problem would obviously depend on the size of $[K]$, the degree of accuracy required etc.; we shall return to this topic in the next section.

Returning to the input data, we can obviously devise a number of ways in which to organize them; one such scheme, which we shall follow throughout, consists of the following sets of cards on which these data will be punched (the scheme given here is also valid for machines with tape input devices):

The 'first' card,
The 'property' cards,
The 'assembly' cards and
The 'stress' cards.

The 'first' card

This card(s) is literally the first card(s) in the input data card deck; the data on this card(s) are not directly concerned with any of the executive operations listed above, but merely instruct the computer on the general parameters of the problem as a whole. Some typical data to be punched on this card(s) are: NN, NEL, NDOF etc. (these are the computer names used here; see Appendix 4 for their meaning).

The 'property' cards

These cards, as their name suggests, contain the geometrical and material properties of elements. When required, the extremity coordinates must also be included. All such details of each element are punched onto a single card (with 'continuation' card(s) if necessary) so that in a given problem the total number of these cards is equal to NEL (see Appendix 4). For example, the property card for the element of Fig. 15.4 will contain the following punched data (not necessarily in this order):

9 .25 3. −3. 1.5 1.5 −1. −1. 2.

which are the respective values of the following computer quantities (see Appendix 4)

I PR T XA XB XC YA YB YC

These cards enable the computer to execute operation 1. If any material (and/or geometrical) property is common to all elements representing the body, then it is convenient to transfer it to the 'first' card. Thus, in a uniform plate

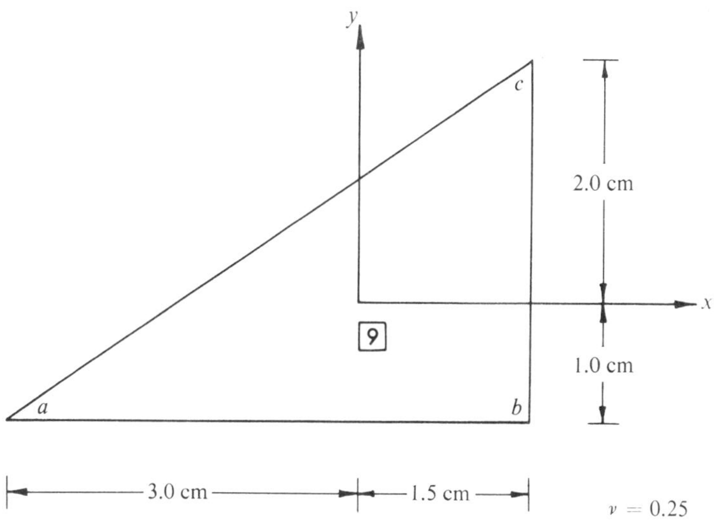

Fig. 15.4 Typical data for a triangular element (thickness = t = 3·0 cm).

problem, for example, it would be more logical to include the constant element thickness in the 'first' card, rather than to repeat the same information in all the property cards.

The 'assembly' cards

These cards enable the computer to assemble (operation 2) the overall stiffness matrix $[K]$ for the entire body by the method of section 3.4. If the extremities $a, b, c \ldots$ (or 1, 2, 3 ... when the extremities are identified by numbers) of element I are attached to the nodes NA,NB,NC, ... respectively, then the assembly card for element I will be punched as

I NEI NA NB NC .. .. ..

where NEI is the total number of extremities of I. In element 2 of Fig. 15.5, for example, I = 2 and NEI = 4; also, since the extremities 1, 2, 3, 4 are respectively attached to the nodes 2, 5, 4, 1, it follows that NA = 2, NB = 5, NC = 4 and ND = 1. The assembly card for this element will therefore be punched as

2 4 2 5 4 1

Next, consider the assembly card for element number 3 (Fig. 15.5). Extremities 1 and 3 of this element are attached to what we shall call 'dummy' nodes, defined as nodes at which all displacements vanish. Consequently, the dummy nodes will not be numbered, as they will be excluded from $[K]$. We shall follow this rule to deal with such elements:

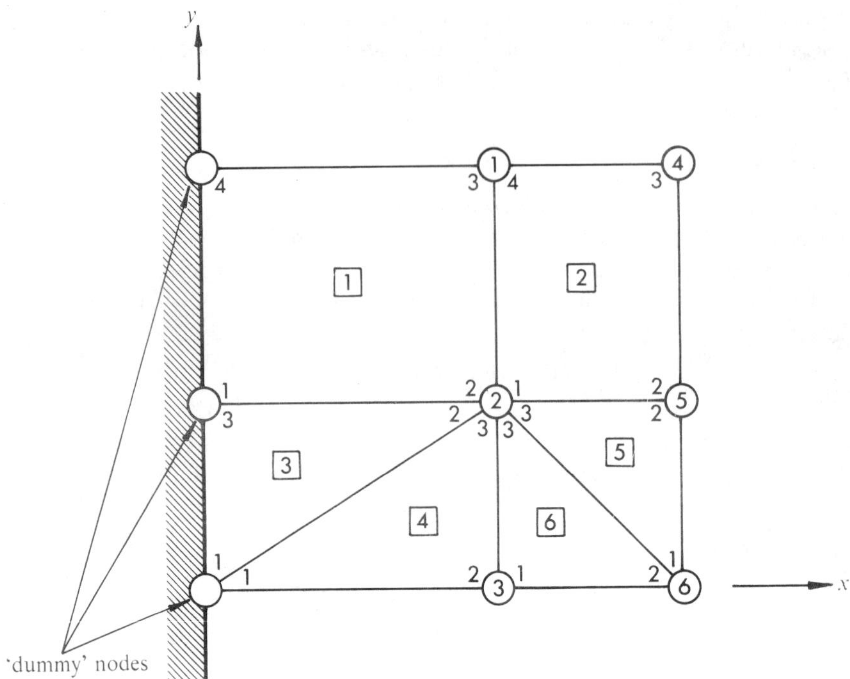

Fig. 15.5 A body subdivided into a combination of rectangular and triangular elements; all displacements vanish on $x = 0$.

If one or more extremities of the element are attached to 'dummy' nodes, then the dummy nodes will be numbered as 'zero'.

Thus, in element 3 NA = NC = 0; also, it is obvious that I = 3, NEI = 3 and NB = 2. The assembly card for this element will therefore be punched as

```
3 3 0 2 0
```

It can be verified that the remaining assembly cards for the finite element model of Fig. 15.5 are:

```
1 4 0 2 1 0
4 3 0 3 2
5 3 6 5 2
6 3 3 6 2
```

Clearly, in a given problem the total number of assembly cards will be equal to NEL. Furthermore, if the problem requires it, these cards can also be used without any alteration to assemble $[m]$ and/or $[\underline{K}]$.

Data cards to deal with the prescribed boundary conditions and to define the overall external force vector $\{P\}$ can also be designed without difficulty.

However, in order not to confuse the beginner, we shall not do this here. Instead, these will be specified in the main program, at the locations indicated. The reader is encouraged to design the data cards to deal with the boundary conditions and $\{P\}$.

The 'stress' cards

The data to be punched in what we have called here 'stress' cards, instruct the computer to calculate the generalized stresses within the elements (operation 4). If in element I the extremities a, b, c, . . . (or 1, 2, 3, . . .) are respectively attached to the nodes NA, NB, NC, . . . , then the following data will be punched on the stress card for element I:

I NEI NA NB NC .. ..

This scheme is valid only when the stresses remain constant throughout element I. If they vary, then the coordinates of the particular point (or points) within the element at which the stresses are to be calculated, must also be included. As an example of this, suppose that we wish to calculate the stresses at points P and Q (Fig. 15.6), which lie within an element the stresses within which vary

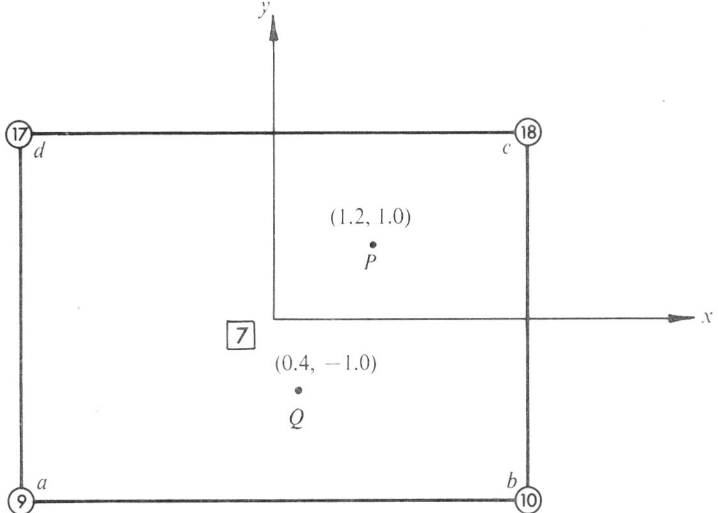

Fig. 15.6 A finite element the stresses within which vary from point to point.

from point to point. Let (XP,YP) and (XQ,YQ) be the respective coordinates of these points. Then, the stress card for this element will be punched as in Fig. 15.7.

I NEI NP NA NB NC ND XP YP XQ YQ

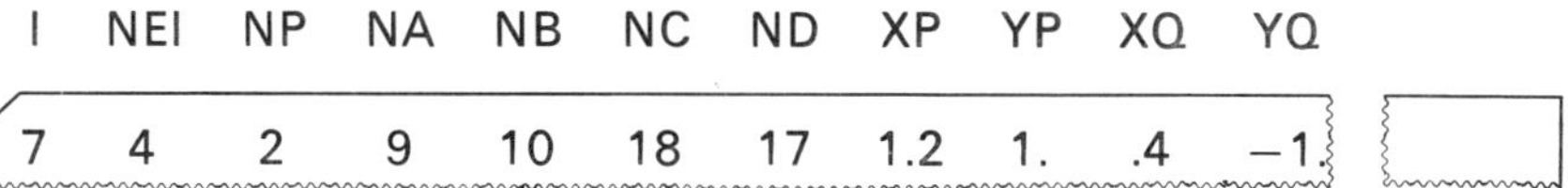

Fig. 15.7 Stress card for the element of Fig. 15.6

NP in Fig. 15.7 is the total number of points within I at which the stresses are to be calculated. The stress card data of each element will be punched onto a single card, with one or more 'continuation' cards if necessary. Thus, in a given problem, the number of 'stress' cards will be NTSC, which is number of elements the stresses within which we wish to calculate.

In these cards, as in the assembly cards, each 'dummy' node will be assigned a zero value.

The following points are worth remembering in connection with the data cards:

(a) The sets of 'first', 'property', 'assembly' and 'stress' cards must be fed into the computer in this order. The relative positions of individual cards within a given set, however, is immaterial (this does not apply to the 'first' card).

(b) In all cards the punched data must be consistent with the input FORMAT's as declared in the main program.

(c) The data organization scheme given here is valid only for the programs of section 15.5.

15.5 Some programs

This section contains three simple computer programs, written in the standard FORTRAN language and tested on the University of London CDC 6600 computer.

As mentioned before, the subroutine(s) required for solving a set of linear simultaneous equations is normally to be found in most program libraries. Indeed, subroutines of various kind are built into the peripheral devices of most large computers. When this facility exists, one has only to CALL the required subroutine(s) from the computer memory whenever necessary. Otherwise, the card deck containing the subroutine(s) must follow the main program deck.

The particular subroutine that will be used here is MSOLVE, which will be given at the end of program 1. It finds the solution to a system of linear simultaneous equations, or inverts a matrix. The method used is that of 'pivotal condensation' with row interchange, followed by back substitution. This subroutine will be called in the main program as

```
CALL MSOLVE(A,B,N,IA,KNOT,OUT)
```

where IA is the dimension of [A] as declared in the main program and OUT the output tape number. When KNOT is equal to 1, the subroutine solves a set of N simultaneous equations of the general form

$$[A]\{x\} = \{B\} \tag{15.14}$$

and the solution $\{x\}$ overwrites $\{B\}$. If KNOT is set equal to 2, on the other hand, then the subroutine forms $[A^{-1}]$ which overwrites $[A]$. $\{B\}$ need not be set if

$[A^{-1}]$ is only required. The subroutine fails if any of the terms on the leading diagonal of $[A]$ is equal to zero.

Program 1

This program is for solving isotropic, uniform or non-uniform flat plate problems by using rectangular elements. The details of this program are as follows:

Operation 1 First, the $[c]$ matrix of Table 10.1 is formed from the extremity coordinates and stored as C(J,K). $[c]$ is then inverted by the $[Z]$ matrix method of section 15.3, by setting $[Z] = [c]^{T}$. The inverse of $[c]$ of element I is stored as A2(I,J,K). The matrix of Table 10.2 is next formed and stored as C(J,K). Matrix $[K_i]$ is then calculated and stored as S(I,J,K). For each element I, its elasticity matrix $[d]$ (Eq. 10.5) is formed and stored as D(I,J,K). This program can also be used to solve orthotropic plate problems merely by making D(I,J,K) equal to the $[d]$ of Eq. (10.27) instead of that of Eq. (10.5).

Operation 2 The overall stiffness matrix is assembled from the assembly cards and is stored as A(J,K). The group of statements lying between and including the statement labels 500 and 501, carry out this assembly. For future reference, we shall call this group the 'assembly block'. This block is universal in that it can assemble the overall stiffness matrix of any discretized body, irrespective of degrees of freedom, number of extremities in the constituent elements, etc.

The method of section 3.4 has been used in programming this block. In order to maintain its universality, the degree of freedom has been retained as the variable NDOF in this block, although program 1 is valid for NDOF = 3 only.

Operation 3 Equation (3.9) for the plate is solved by MSOLVE and the solution, stored as P(J), printed out.

Operation 4 The $[N]$ matrix of Eq. (10.13) is formed and stored as A(J,K) for each point within the element at which the stress-moments are to be calculated; these are then computed from Eq. (10.18) for each specified point within the element and the results printed out. This process is repeated for all elements whose stress cards are supplied.

Note that in this program the centroid of each element is taken as the origin of its local coordinates and that in every element extremity a lies in the 3rd quadrant, as in Fig. 15.6.

As an example, let us now apply this program to solve the problem of Fig. 10.6; consider, for instance, the model of Fig. 10.6b whose magnified version is given in Fig. 15.8. The vectors of Eq. (10.16) and (10.17) in this case are

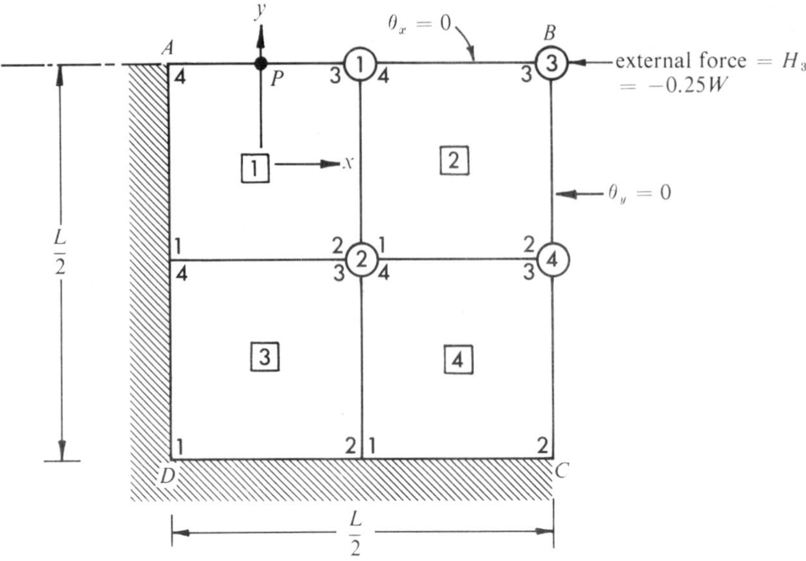

Fig. 15.8 Magnified model (b) of Fig. 10.6. All displacements vanish on *AD* and *DC* ('dummy' nodes not shown).

respectively

$$\{P\} = \{\underset{1}{H_1}\ \underset{2}{M_{x1}}\ \underset{3}{M_{y1}}\ \underset{4}{H_2}\ \underset{5}{M_{x2}}\ \underset{6}{M_{y2}}\ \underset{7}{H_3}\ \underset{8}{M_{x3}}\ \underset{9}{M_{y3}}\ \underset{10}{H_4}\ \underset{11}{M_{x4}}\ \underset{12}{M_{y4}}\}^{\mathrm{T}}$$

and

$$\{\delta\} = \{\underset{1}{w_1}\ \underset{2}{\theta_{x1}}\ \underset{3}{\theta_{y1}}\ \underset{4}{w_2}\ \underset{5}{\theta_{x2}}\ \underset{6}{\theta_{y2}}\ \underset{7}{w_3}\ \underset{8}{\theta_{x3}}\ \underset{9}{\theta_{y3}}\ \underset{10}{w_4}\ \underset{11}{\theta_{x4}}\ \underset{12}{\theta_{y4}}\}^{\mathrm{T}}$$

The applied external force in this case is simply $H_3 = -0{\cdot}25$ (taking $W = 1$). Consequently, since this program recognizes $\{P\}$ as P(J), it is clear that within the range J=1,12 all values of P(J) will be zero except P(7) = $-0{\cdot}25$.

As for the boundary conditions, it is obvious that in this case

$$\underset{2}{\theta_{x1}} = \underset{8}{\theta_{x3}} = \underset{9}{\theta_{y3}} = \underset{12}{\theta_{y4}} = 0$$

One of the easiest ways* of satisfying a zero displacement condition of this type is by replacing the leading diagonal term of $[K]$, corresponding to that displacement, by a very large number. Since the computer recognizes $[K]$ as A(J,K), the above conditions will therefore be satisfied when we introduce the

* This is an alternative to method 1 (section 5.1) and is valid when $\overline{U} = 0$.

following statements in the main program at the appropriate location:

```
XX = 1.0E+30
A(2,2)   = XX
A(8,8)   = XX
A(9,9)   = XX
A(12,12) = XX
```

With reference to the discussion of data cards in the previous section, the following punched data will be required for this problem:

Property cards (Format declared in label 101)

I	XA	YA	T	PR	E
1	−.125	−.125	1.0	.25	1.0
2	−.125	−.125	1.0	.25	1.0
3	−.125	−.125	1.0	.25	1.0
4	−.125	−.125	1.0	.25	1.0

Because the data in every card are the same, it would be more logical in this case to enter them in the 'first' card rather than repeating them in all the property cards.

Assembly cards (Format declared in label 202)

I	NEI	NA	NB	NC	ND
1	4	0	2	1	0
2	4	2	4	3	1
3	4	0	0	2	0
4	4	0	0	4	2

Stress cards (Format declared in label 402)

The total number of these cards is equal to the number of elements the stresses within which we wish to calculate. Suppose that we wish to calculate the stresses at the single point P (Fig. 15.8) in element 1. Then, NP=1; also, since the coordinates of P are (0, 0·125), the required data card is

I	NEI	NP	NA	NB	NC	ND	XP	YP
1	4	1	0	2	1	0	0.	.125

The first card (Format declared in label 100)

In this case NN = 4, NDOF = 3, NEL = 4 and since we are interested in the stress within a single element, NTSC = 1. Consequently, the 'first' card will be

punched as

NN	NDOF	NEL	NTSC
4	3	4	1

Program 2

This program, valid for NDOF=2, solves plane isotropic elasticity problems by using triangular elements. It solves a plane strain problem when the first card parameter MP is set equal to 1, and a plane stress problem when this parameter is set equal to 2. Both the prescribed boundary conditions and the external force vector $\{P\}$, which the computer recognizes as P(J), must be introduced at the indicated locations in the main program. The main features of this program are as follows:

Operation 1 The property data are read in; the matrix $[f] = [N][c^{-1}]$ is formed from Table 11.1 and Eq. (11.4b), and the $[f]$ matrix of element I is stored as A2(I,J,K). The $[d]$ matrix, defined by Eq. (8.18) for plane strain and by Eq. (8.19) for plane stress, is formed for each element and stored as D(I,J,K). Finally, $[K_i]$ is formed from Eq. (11.5) and stored as S(I,J,K).

Operation 2 The 'assembly block' of program 1 (see operation 2, program 1) to be added here and also the prescribed boundary conditions and the vector $\{P\}$.

Operation 3 Equation (3.9) for the problem is solved by MSOLVE and the computed components of displacement printed out.

Operation 4 The stresses within individual elements are computed from Eq. (11.10) and the results printed out (initial and thermal strains not considered).

The various input data to be supplied are as follows:

Card(s)	Data	To be punched according to FORMAT declared in label
'first'	NN,NDOF,NEL,NTSC,MP	100
'property'	I,XA,XB,XC,YA,YB,YC,G,PR,T*	101
'assembly'	I,NEI,NA,NB,NC	202 (program 1)
'stress'	I,NEI,NA,NB,NC	401

* In the case of plane strain set T=1.0 in all property cards.

Program 3

This program solves the two dimensional anisotropic Poisson equation (Eq. 13.2.3) by using triangular 'field' elements shown in Fig. 13.6a. It can

also be used to solve the two dimensional anisotropic Laplace equation (Eq. 13.2.39) merely by setting FI = 0.0 in every property card. The details of this program are as follows:

Operation 1 The property card data are read in; the quantity $\Delta I/3$ of Eq. (13.2.22) is calculated for every element and is stored as F(I). The $[c]$ matrix, defined by Eq. (13.2.16), is formed and subsequently the $[K_i]$ matrix of each element is calculated from Eq. (13.2.23); the $[K_i]$ of element I is stored as S(I,J,K).

Operation 2 The $[K]$ matrix, stored as A(J,K), for the entire body is assembled from the assembly card data. Since NDOF = 1 here, these statements are simpler than those of the 'assembly block' of program 1. (We could have used the assembly block to assemble $[K]$, if we wanted to, but this would have been marginally wasteful). The body force contributions to $\{P\}$, which the program recognizes as P(J), are also calculated.

Operation 3 Equation (13.2.30) for the problem is solved by MSOLVE and the values of PHI(ϕ) printed out.

The various input data to be supplied are as follows:

Card(s)	Data	To be punched according to FORMAT declared in label
'first'	NN,NEL	100
'property'	I,XA,YA,XB,YB,XC,YC,RX,RY,FI*	101
'assembly'	I,NEI,NA,NB,NC	201

* For using this program to solve the two dimensional Laplace equation, set FI = 0.0 in every property card.

Operation 4, which is absent in this program, is easily written to determine the 'stresses' (Eq. 13.2.6) within individual elements. This is left to the reader as an exercise.

```
C     PROGRAM 1; THIS PROGRAM IS FOR SOLVING FLAT PLATE PROBLEMS BY
C     USING RECTANGULAR ELEMENTS. CENTROID OF ELEMENT TAKEN AS ORIGIN OF
C     COORDINATES. VECTOR P AND BOUNDARY CONDITIONS TO BE INSERTED AT
C     LOCATIONS INDICATED. REFER TO FIG. 10.2.
      DIMENSION A(100,100),Y(100),P(150),C(12,12),Z(12,12),A1(12,12)
      DIMENSION A2(20,12,12),S(20,12,12),X(12),XP(5),YP(5)
      DIMENSION D(20,3,3),N1(20),N2(20),N3(20)
      READ(5,100) NN,NDOF,NEL,NTSC
100   FORMAT(4I3)
C     ****** OPERATION 1 BEGINS HERE **********
      WRITE(6,900)
900   FORMAT(//10X,18HPROPERTY CARDS****,10X,1HI,8X,2HXA,8X,2HYA,9X,1HT,
     18X,2HPR,9X,1HE/)
      IK=1
112   READ(5,101) I,XA,YA,T,PR,E
```

```
101   FORMAT(I3,5F10.0)
      WRITE(6,901) I,XA,YA,T,PR,E
901   FORMAT(28X,I11,5F10.4)
      X(1)  =  XA
      X(4)  = -XA
      X(7)  = -XA
      X(10) =  XA
      Y(1)  =  YA
      Y(4)  =  YA
      Y(7)  = -YA
      Y(10) = -YA
      DO 102 J=1,12
      DO 102 K=1,12
      C(J,K)  = 0.0
      A1(J,K)=0.0
102   A2(I,J,K)=0.0
      DO 103 J=1,10,3
      C(J,1)  = 1.0
      C(J,2)  = X(J)
      C(J,3)  = Y(J)
      C(J,4)  = X(J)**2
      C(J,5)  = X(J)*Y(J)
      C(J,6)  = Y(J)**2
      C(J,7)  = X(J)**3
      C(J,8)  = Y(J)*X(J)**2
      C(J,9)  = X(J)*Y(J)**2
      C(J,10)= Y(J)**3
      C(J,11)= Y(J)*X(J)**3
      C(J,12)= X(J)*Y(J)**3
      C(J+1,3)   =   1.0
      C(J+1,5)   =   X(J)
      C(J+1,6)   =   2.*Y(J)
      C(J+1,8)   =   X(J)**2
      C(J+1,9)   =   2.*X(J)*Y(J)
      C(J+1,10)  =   3.*Y(J)**2
      C(J+1,11)  =   X(J)**3
      C(J+1,12)  =   3.*X(J)*Y(J)**2
      C(J+2,2)   = -1.0
      C(J+2,4)   = -2.*X(J)
      C(J+2,5)   = -Y(J)
      C(J+2,7)   = -3.*X(J)**2
      C(J+2,8)   = -2.*X(J)*Y(J)
      C(J+2,9)   = -Y(J)**2
      C(J+2,11)  = -3.*Y(J)*X(J)**2
103   C(J+2,12)  = -Y(J)**3
      DO 104 J=1,12
      DO 104 K=1,12
      A(J,K)=0.0
104   Z(J,K)=C(K,J)
      DO 105 N=1,12
      DO 105 J=1,12
      DO 105 K=1,12
105   A1(J,N)= A1(J,N) + C(J,K)*Z(K,N)
      CALL MSOLVE(A1,Y,12,12,2,6)
      DO 106 N=1,12
      DO 106 J=1,12
      C(N,J)=0.0
      DO 106 K=1,12
106   A2(I,J,N) = A2(I,J,N) + Z(J,K)*A1(K,N)
      HX= -XA
      HY= -YA
      AREA= 4.*HX*HY
      Q = (1.0-PR)/2.0
      C(4,4)   = 4.0
```

```
      C(4,6)   = 4.*PR
      C(5,5)   = 4.*Q
      C(5,11)  = 4.*Q*HX**2
      C(5,12)  = 4.*Q*HY**2
      C(6,6)   = 4.0
      C(7,7)   = 12.*HX**2
      C(7,9)   = 4.*PR*HX**2
      C(8,8)   = 4.*(HY**2 + 4.*Q*HX**2)/3.0
      C(8,10)  = 4.*PR*HY**2
      C(9,9)   = 4.*(HX**2 + 4.*Q*HY**2)/3.0
      C(10,10)= 12.*HY**2
      C(11,11)= AREA**2/4.0 + 36.*Q*HX**4/5.0
      C(11,12)= (PR+Q)*AREA**2/4.0
      C(12,12)= AREA**2/4.0 + 36.*Q*HY**4/5.0
      DO 107 J=1,11
      NX=J+1
      DO 107 K=NX,12
107   C(K,J)=C(J,K)
      DO 108 N=1,12
      DO 108 J=1,12
      A1(N,J)=0.0
      DO 108 K=1,12
108   A(J,N)= A(J,N) + C(J,K)*A2(I,K,N)
      DO 109 N=1,12
      DO 109  J=1,12
      DO 109  K=1,12
109   A1(J,N)   = A1(J,N) + A2(I,K,J)*A(K,N)
      D1= E*T**3/(12.*(1.0-PR**2))
      AR= AREA*D1
      DO 110 J=1,12
      DO 110 K=1,12
110   S(I,J,K)= A1(J,K)*AR
      DO 111  J=1,3
      DO 111  K=1,3
111   D(I,J,K)=0.0
      D(I,1,1)= D1
      D(I,1,2)= D1*PR
      D(I,2,1) = D1*PR
      D(I,2,2) = D1
      D(I,3,3) = D1*Q
      IK=IK+1
      IF(IK.LE.NEL) GO TO 112
C     OPERATION 1 ENDS AND OPERATION 2 BEGINS HERE *********
      WRITE(6,902)
902   FORMAT(//10X,19HASSEMBLY CARDS*****,10X,1HI,2X,3HNEI,5X,5HNODES/)
500   NN1=NN*NDOF
      DO 200 J=1,NN1
      P(J)=0.0
      DO 200 K=1,NN1
200   A(J,K)=0.0
      DO 201 J=1,12
201   N2(J)=NDOF*(J-1)
      NDUMMY=NN1+1
      IK=1
207   READ(5,202)I,NEI,(N1(J),J=1,NEI)
202   FORMAT(2I3,3X,20I3)
      WRITE(6,903) I,NEI,(N1(J),J=1,NEI)
903   FORMAT(29X,I11,I5,5X,20I3)
      DO 203 J=1,NEI
      IF(N1(J).EQ.0) N1(J)=NDUMMY
      IF(N1(J).NE.NDUMMY) N1(J)=NDOF*(N1(J)-1)
203   CONTINUE
      DO 204 J=1,NEI
      IF(N1(J).EQ.NDUMMY) GO TO 204
```

```
      NJ=N1(J)
      NL=N2(J)
      DO 205 K=1,NEI
      IF(N1(K).EQ.NDUMMY) GO TO 205
      NK=N1(K)
      NM=N2(K)
      DO 206 L=1,NDOF
      DO 206 M=1,NDOF
206   A(NJ+L,NK+M)= A(NJ+L,NK+M) + S(I,NL+L,NM+M)
205   CONTINUE
204   CONTINUE
      IK=IK+1
501   IF(IK.LE.NEL) GO TO 207
C     ****** DEFINE VECTOR P AND PRESCRIBED BOUNDARY CONDITIONS HERE *
C     OPERATION 2 ENDS AND OPERATION 3 BEGINS HERE***********
      CALL MSOLVE(A,P,NN1,100,1,6)
      WRITE(6,904)
904   FORMAT(//10X,4HNODE,5X,8HVERT DFL,10X,6HTHETAX,10X,6HTHETAY/)
      DO 300 J=1,NN
      K=NDOF*(J-1)+1
300   WRITE(6,905) J,P(K),P(K+1),P(K+2)
905   FORMAT(10X,I4,F13.7,2F16.7)
C     OPERATION 3 ENDS AND OPERATION 4 BEGINS HERE***********
      WRITE(6,906)
906   FORMAT(//10X,14HSTRESS MOMENTS,5X,1HI,5X,2HXP,5X,2HYP,9X,2HMX,9X,2
     1HMY,8X,3HMYX/)
      DO 400 J=1,3
      DO 400 K=1,12
400   A(J,K)=0.0
      NDUMMY = NN+1
      DO 401 J=1,NDOF
401   P(NDOF*NDUMMY-J+1)=0.0
      A(1,4) = -2.0
      A(2,6) = -2.0
      A(3,5) = -2.0
      IK=1
410   READ(5,402) I,NEI,NP,(N3(J),J=1,NEI),(XP(K),YP(K),K=1,NP)
402   FORMAT(3I3,2X,10I3,6F5.0)
      L=0
      DO 403 K=1,NEI
      N=N3(K)
      IF(N.EQ.0) N=NDUMMY
      DO 404 J=1,3
404   Y1(J+L)=P(3*N-3+J)
      L = L+3
403   CONTINUE
      DO 405 J=1,12
      X(J)=0.0
      DO 405 K=1,12
405   X(J)=X(J) + A2(I,J,K)*Y1(K)
      DO 406 N=1,NP
      DO 407 J=1,3
      Y(J)=0.0
407   Y1(J)=0.0
      A(1,7)  = -6.*XP(N)
      A(1,8)  = -2.*YP(N)
      A(1,11)= -6.*XP(N)*YP(N)
      A(2,9)  = -2.*XP(N)
      A(2,10)= -6.*YP(N)
      A(2,12)= -6.*XP(N)*YP(N)
      A(3,8)  = -4.*XP(N)
      A(3,9)  = -4.*YP(N)
      A(3,11)= -6.*XP(N)**2
      A(3,12)= -6.*YP(N)**2
      DO 408 J=1,3
```

```
      DO 408 K=1,12
408   Y1(J)= Y1(J) + A(J,K)*X(K)
      DO 409 J=1,3
      DO 409 K=1,3
409   Y(J) = Y(J) + D(I,J,K)*Y1(K)
406   WRITE(6,907) I,XP(N),YP(N),(Y(J),J=1,3)
907   FORMAT(24X,I6,2F7.3,3F11.5)
      IK=IK+1
      IF(IK.LE.NTSC) GO TO 410
C     MAIN PROGRAM ENDS HERE: ADD SUBROUTINE MSOLVE, FOLLOWED BY THE
C     DATA CARDS.
      STOP
      END
```

```
      SUBROUTINE MSOLVE (A,B,N,IA,KNOT,OUT)
      DIMENSION A(IA,N),B(N),C(400),IND(400)
      INTEGER OUT
      M=N
100   AMAX=0.0
      DO 2 I=1,M
      IND(I)=I
      IF(ABS(A(I,1)) - AMAX) 2,2,3
3     AMAX=ABS(A(I,1))
      I4=I
2     CONTINUE
      MM=M-1
      DO 111 J=1,MM
      IF(I4-J) 6,6,4
4     ISTO=IND(J)
      IND(J)=IND(I4)
      IND(I4)=ISTO
      DO 5 K=1,M
      STO=A(I4,K)
      A(I4,K)=A(J,K)
      A(J,K)=STO
5     CONTINUE
      GO TO (7,6),KNOT
7     STO=B(I4)
      B(I4)=B(J)
      B(J)=STO
6     AMAX=0.0
      J1=J+1
      IF(A(J,J).EQ.0.0) WRITE(OUT,92) J
92    FORMAT(//5X,34HMSOLVE ERROR-THE DIAGONAL ON ROW I3,
      9H IS ZERO.//)
      DO 11 I=J1,M
      A(I,J)=A(I,J)/A(J,J)
      DO 10 K=J1,M
      A(I,K)=A(I,K)- A(I,J)*A(J,K)
      IF(K - J1) 14,14,10
14    IF(ABS(A(I,K)) - AMAX) 10,10,17
17    AMAX=ABS(A(I,K))
      I4=I
10    CONTINUE
9     GO TO (12,11),KNOT
12    B(I)=B(I) - A(I,J)*B(J)
11    CONTINUE
111   CONTINUE
      IF(A(M,M)) 94,93,94
93    WRITE(OUT,92) M
```

```
      STOP
94    GO TO (19,18),KNOT
19    DO 127 I1=1,M
      I=M+1-I1
      IF(M - I) 327,327,28
28    I2=I+1
      DO 32 K=I2,M
      B(I)=B(I) - A(I,K)*B(K)
32    CONTINUE
327   B(I)=B(I)/A(I,I)
127   CONTINUE
18    GO TO (64,65),KNOT
65    DO 140 I1=1,MM
      I=M+1-I1
      I2=I-1
      DO 41 J1=1,I2
      J=I2+1-J1
      J2=J+1
      W1= -A(I,J)
      IF(I2 - J2) 141,43,43
43    DO 42 K=J2,I2
      W1= W1 - A(K,J)*C(K)
42    CONTINUE
141   C(J)=W1
41    CONTINUE
      DO 40 K=1,I2
      A(I,K)=C(K)
40    CONTINUE
140   CONTINUE
      DO 150 I1=1,M
      I=M+1-I1
      I2=I+1
      W=A(I,I)
      IF(W) 95,96,95
96    WRITE(OUT,92) I
      STOP
95    DO 56 J=1,M
      IF(I - J) 52,53,54
52    W1=0.0
      GO TO 55
53    W1=1.0
      GO TO 55
54    W1=A(I,J)
55    IF(I1 - 1) 156,156,57
57    DO 58 K=I2,M
      W1=W1 - A(I,K)*A(K,J)
58    CONTINUE
156   C(J)=W1
56    CONTINUE
      DO 50 J=1,M
      A(I,J)=C(J)/W
50    CONTINUE
150   CONTINUE
      DO 60 I=1,M
63    IF(IND(I) - I) 61,60,61
61    J=IND(I)
      DO 62 K=1,M
      STO=A(K,I)
      A(K,I)=A(K,J)
      A(K,J)=STO
62    CONTINUE
      ISTO=IND(J)
      IND(J)=J
      IND(I)=ISTO
```

```
      GO TO 63
60    CONTINUE
64    RETURN
      END
```

```
C     PROGRAM 2; THIS PROGRAM IS FOR SOLVING PLANE ISOTROPIC ELASTIC
C     PROBLEMS BY USING TRIANGULAR ELEMENTS: CENTROID OF ELEMENT TAKEN
C     AS ORIGIN OF COORDINATES. VECTOR P AND THE BOUNDARY CONDITIONS TO
C     BE DEFINED AT THE LOCATIONS INDICATED. REFER TO SECTION 11.1. SET
C     MP=1 FOR PLANE STRAIN AND MP=2 FOR PLANE STRESS IN THE FIRST CARD.
      DIMENSION A(100,100),P(150),C(6,6),Z(6,6),S(20,6,6),A2(20,6,6)
      DIMENSION D(20,3,3),N1(20),N2(20),N3(20),X(6),Y1(6)
      READ(5,100) NN,NDOF,NEL,NTSC,MP
100   FORMAT(5I3)
C     OPERATION 1 BEGINS HERE**************
      WRITE(6,900)
900   FORMAT(//10X,18HPROPERTY CARDS****,10X,1HI,8X,2HXA,8X,2HYA,8X,2HXB
     1,8X,2HYB,8X,2HXC,8X,2HYC,9X,1HG,3X,2HPR/)
      IK=1
108   READ(5,101) I,XA,YA,XB,YB,XC,YC,G,PR,T
101   FORMAT(I3,6F10.0/3F10.0)
      WRITE(6,901) I,XA,YA,XB,YB,XC,YC,G,PR
901   FORMAT(28X,I11,6F10.5,F10.2,F5.3)
      DO 102 J=1,6
      DO 102 K=1,6
      A(J,K)   = 0.0
      C(J,K)   = 0.0
      S(I,J,K) = 0.0
102   A2(I,J,K)= 0.0
      A2(I,1,1) = YB-YC
      A2(I,1,3) = YC-YA
      A2(I,1,5) = YA-YB
      A2(I,2,2) = XC-XB
      A2(I,2,4) = XA-XC
      A2(I,2,6) = XB-XA
      A2(I,3,1) = XC-XB
      A2(I,3,2) = YB-YC
      A2(I,3,3) = XA-XC
      A2(I,3,4) = YC-YA
      A2(I,3,5) = XB-XA
      A2(I,3,6) = YA-YB
      AREA= ((YB-YC)*(XA-XB) - (YA-YB)*(XB-XC))/2.0
      AR=2.*AREA
      DO 103 J=1,3
      DO 103 K=1,6
103   A2(I,J,K) = A2(I,J,K)/AR
      DO 104 J=1,3
      DO 104 K=1,3
104   D(I,J,K)=0.0
      IF(MP.EQ.2) GO TO 500
      D(I,1,1) = 2.*G*(1.0-PR)/(1.0-2.*PR)
      D(I,1,2) = D(I,1,1)*PR/(1.0-PR)
      D(I,2,1) = D(I,1,2)
      D(I,2,2) = D(I,1,1)
      D(I,3,3) = G
      IF(MP.EQ.1) GO TO 501
500   D(I,1,1) = 2.*G/(1.0-PR)
      D(I,1,2) = D(I,1,1)*PR
      D(I,2,1) = D(I,1,2)
      D(I,2,2) = D(I,1,1)
```

```
      D(I,3,3) = G
501   DO 105 N=1,6
      DO 105 J=1,3
      DO 105 K=1,3
105   C(J,N) = C(J,N) + D(I,J,K)*A2(I,K,N)
      DO 106 N=1,6
      DO 106 J=1,6
      DO 106 K=1,3
106   S(I,J,N) = S(I,J,N) + A2(I,K,J)*C(K,N)
      DO 107 J=1,6
      DO 107 K=1,6
107   S(I,J,K) = S(I,J,K)*AREA*T
      IK=IK+1
      IF(IK.LE.NEL) GO TO 108
C     OPERATION 1 ENDS AND OPERATION 2 BEGINS HERE*******
C     ADD 'ASSEMBLY BLOCK' OF PROGRAM 1 HERE.
C     DEFINE VECTOR P AND THE BOUNDARY CONDITIONS HERE.
C     OPERATION 2 ENDS AND OPERATION 3 BEGINS HERE.
      CALL MSOLVE(A,P,NN1,100,1,6)
      WRITE(6,904)
904   FORMAT(//10X,4HNODE,5X,10HHORZ DISPL,5X,10HVERT DISPL/)
      DO 300 J=1,NN
      K=2*(J-1)+1
300   WRITE(6,905) J,P(K),P(K+1)
905   FORMAT(10X,I4,2F15.6)
C     OPERATION 3 ENDS AND OPERATION 4 BEGINS HERE*******
      WRITE(6,906)
906   FORMAT(/10X,14HSTRESSES******,5X,1HI,6X,3HSXX,6X,3HSYY,6X,3HSXY/)
      NDUMMY = NN+1
      DO 400 J=1,NDOF
400   P(NDOF*NDUMMY-J+1)=0.0
      IK=1
406   READ(5,401) I,NEI,(N3(J),J=1,NEI)
401   FORMAT(2I3,2X,20I3)
      DO 402 N=1,6
      X(N)=0.0
      DO 402 J=1,3
      C(J,N)=0.0
      DO 402 K=1,3
402   C(J,N)= C(J,N) + D(I,J,K)*A2(I,K,N)
      L = 0
      DO 403 K=1,NEI
      N = N3(K)
      IF(N.EQ.0) N=NDUMMY
      DO 404 J=1,2
404   Y1(J+L)=P(2*N-2+J)
      L=L+2
403   CONTINUE
      DO 405 J=1,3
      DO 405 K=1,6
405   X(J) = X(J) + C(J,K)*Y1(K)
      WRITE(6,907) I,X(1),X(2),X(3)
907   FORMAT(24X,I6,3F9.5)
      IK=IK+1
      IF(IK.LE.NTSC) GO TO 406
C     MAIN PROGRAM ENDS HERE. ADD MSOLVE FOLLOWED BY THE DATA CARDS.
      STOP
      END
```

```
C     PROGRAM 3; THIS PROGRAM SOLVES THE POISSON EQUATION IN TWO DIMEN-
C     SIONS BY USING TRIANGULAR ELEMENTS. CENTROID OF ELEMENT TAKEN AS
```

```
C     ORIGIN OF COORDINATES. VECTOR P AND BOUNDARY CONDITIONS TO BE
C     DEFINED AT LOCATIONS INDICATED. OPERATION 4 IS NOT INCLUDED.
C     THIS PROGRAM WILL SOLVE THE LAPLACE EQUATION IN TWO DIMENSIONS
C     WHEN WE SET F(I)=0.0 IN EVERY PROPERTY CARD.
      DIMENSION A(100,100),P(100),S(20,3,3),F(100),C(3,3),A1(3,3),N1(20)
      READ(5,100) NN,NEL
100   FORMAT(2I3)
C     OPERATION 1 BEGINS HERE************************
      WRITE(6,900)
900   FORMAT(//10X,18HPROPERT CARDS****,10X,1HI,6X,2HXA,6X,2HYA,6X,2HXB,
     16X,2HYB,6X,2HXC,6X,2HYC,6X,2HRX,6X,2HRY,6X,2HFI/)
      IK=1
105   READ(5,101) I,XA,YA,XB,YB,XC,YC,RX,RY,FI
101   FORMAT(I3,6F10.0/3F10.0)
      WRITE(6,901) I,XA,YA,XB,YB,XC,YC,RX,RY,FI
901   FORMAT(28X,I11,9F8.4)
      DO 102 J=1,3
      DO 102 K=1,3
102   A(J,K)=0.0
      AREA = ((YB-YC)*(XA-XB) - (YA-YB)*(XB-XC))/2.0
      F(I) = FI*AREA/3.0
      A(2,2)= AREA/RX
      A(3,3)= AREA/RY
      C(1,1)= 1.0
      C(1,2)= XA
      C(1,3)= YA
      C(2,1)= 1.0
      C(2,2)= XB
      C(2,3)= YB
      C(3,1)= 1.0
      C(3,2)= XC
      C(3,3)= YC
      CALL MSOLVE(C,P,3,3,2,6)
      DO 103 N=1,3
      DO 103 J=1,3
      A1(J,N) = 0.0
      DO 103 K=1,3
103   A1(J,N) = A1(J,N) + A(J,K)*C(K,N)
      DO 104 N=1,3
      DO 104 J=1,3
      S(I,J,N) = 0.0
      DO 104 K=1,3
104   S(I,J,N) = S(I,J,N) + C(K,J)*A1(K,N)
      IK=IK+1
      IF(IK.LE.NEL) GO TO 105
C     OPERATION 1 ENDS AND OPERATION 2 BEGINS HERE**********
      WRITE(6,902)
902   FORMAT(//10X,19HASSEMBLY CARDS*****,10X,1HI,2X,3HNEI,5X,5HNODES/)
      DO 200 J=1,NN
      P(J)=0.0
      DO 200 K=1,NN
200   A(J,K) = 0.0
C     DEFINE THE EXTERNAL FORCE VECTOR P HERE***********
      IK=1
204   READ(5,201) I,NEI,(N1(J),J=1,NEI)
201   FORMAT(2I3,3X,20I3)
      WRITE(6,903) I,NEI,(N1(J),J=1,NEI)
903   FORMAT(29X,I11,I5,5X,20I3)
      DO 202 J=1,NEI
      IF(N1(J).EQ.0) GO TO 202
      NJ=N1(J)
      P(NJ)=P(NJ) + F(I)
      DO 203 K=1,NEI
      IF(N1(K).EQ.0) GO TO 203
```

```
      NK=N1(K)
      A(NJ,NK) = A(NJ,NK) + S(I,J,K)
203   CONTINUE
202   CONTINUE
      IK=IK+1
      IF(IK.LE.NEL) GO TO 204
C     DEFINE PRESCRIBED BOUNDARY CONDITIONS HERE************
C     OPERATION 2 ENDS AND OPERATION 3 BEGINS HERE**********
      CALL MSOLVE(A,P,NN,100,1,6)
      WRITE(6,904) (J,P(J),J=1,NN)
904   FORMAT(/10X,4HNODE,13HVALUES OF PHI/(I14,F18.6))
C     MAIN PROGRAM ENDS HERE. ADD SUBROUTINE MSOLVE FOLLOWED BY THE
C     INPUT DATA CARDS.
      STOP
      END
```

15.6 Solution with $[K]$ as a rectangular matrix

In the foregoing programs $[K]$ was assembled and stored in the computer as a square matrix. However, since $[K^{-1}]$ is usually not required in the majority of problems, $[K]$ can be alternatively assembled and stored as a much smaller rectangular matrix. We are able to do this because, if the nodes of the discretized body are numbered in a systematic way, then the non-zero terms of $[K]$ will normally occur only in the vicinity of its leading diagonal. Consider, for instance, the $[K]$ matrix shown in Fig. 15.9a; the so called 'band width' of $[K]$ here is equal to W. Since the order of $[K]$ is N, a total of N^2 'words' would

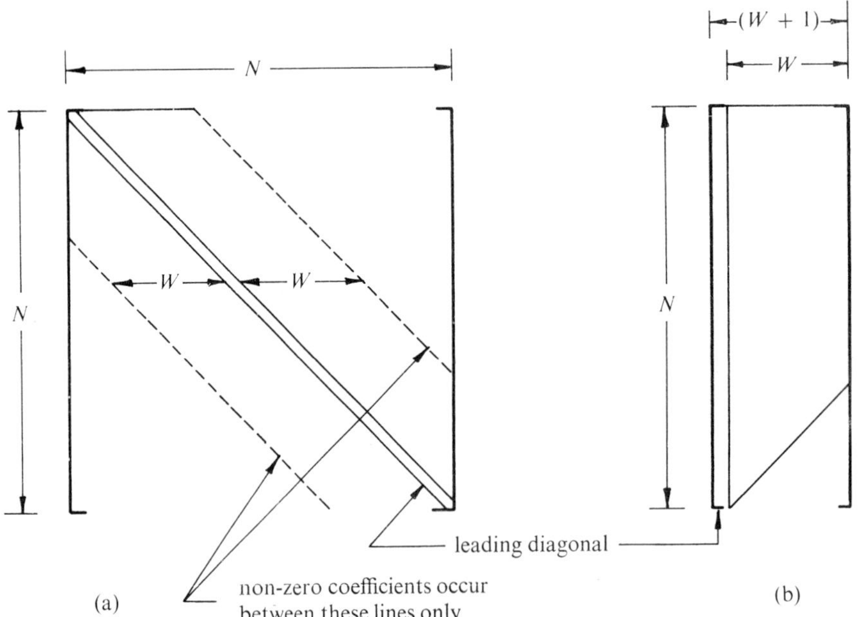

Fig. 15.9 (a) The $[K]$ matrix in the square form. (b) The $[K]$ matrix in the rectangular form; symmetrical half only.

obviously be required to store it in the computer as a square matrix, as in Fig. 15.9a. If, on the other hand, we take advantage of the symmetry of $[K]$ and assemble it as a rectangular matrix of N rows and $(W + 1)$ columns, as in Fig. 15.9b, then only $N(W + 1)$ 'words' would be required to store it. Obviously, a considerable saving in computer storage requirement is to be made in this way, particularly in large problems, by minimizing the 'band width' W.

In a given problem the value of W will be determined by the way the nodes of the discretized body are numbered. The maximum saving to be achieved by 'banding' the $[K]$ matrix in this way, will obviously not materialize if the node numbering is haphazard. If, on the other hand, the numbering is carried out in a systematic and sequential manner, then W and hence the storage requirement would be minimized. Indeed, a minimization problem of this type is a very interesting one in its own right and the reader is encouraged, when dealing with large problems, to try out two or more different schemes of node numbering to determine the minimum value of W.

Quite apart from the much reduced storage requirement, the computer processing time for the 'rectangularized' or 'banded' $[K]$ matrix is also considerably less than when $[K]$ is square. Thus, on both counts, the use of the rectangular $[K]$ matrix is a very attractive proposition and is strongly recommended. Standard subroutines are usually available in program libraries to solve Eq. (15.14) in which $[A]$ is a 'band' matrix of the type just described.

As an example, consider the following matrix:

$$\begin{bmatrix} 24 & 0 & -12 & 6 & 0 & 0 \\ 0 & 8 & -6 & 2 & 0 & 0 \\ -12 & -6 & 24 & 0 & -12 & 6 \\ 6 & 2 & 0 & 8 & -6 & 2 \\ 0 & 0 & -12 & -6 & 12 & -6 \\ 0 & 0 & 6 & 2 & -6 & 4 \end{bmatrix}$$

When the coefficients of this symmetric matrix are reorganized according to the scheme of Fig. 15.9b, the following rectangular or band matrix results:

$$\begin{bmatrix} 24 & 0 & -12 & 6 \\ 8 & -6 & 2 & 0 \\ 24 & 0 & -12 & 6 \\ 8 & -6 & 2 & 0 \\ 12 & -6 & 0 & 0 \\ 4 & 0 & 0 & 0 \end{bmatrix}$$

Clearly, $W = 3$ here and the leading diagonal of the square matrix has become the first column of the band matrix. It follows therefore that when a unsymmetrical square matrix is written as a band matrix, its leading diagonal would become the $(W + 1)$-th column of the band matrix; also, the size of the band matrix in this case would be $N(2W + 1)$.

Appendix 1 Conversion of some common SI units to equivalent F.P.S units

Length	1 m	= 3·28083 ft
Area	1 m^2	= 10·76391 ft^2
Mass	1 kg	= 2·20462 lb
Density	1 kg/m^3	= 0·062428 lb/ft^3
Force	1 N	= 0·224809 lbf
Moment	1 Nm	= 0·737562 lbf ft
Stress	1 MN/m^2	= 145·038 lbf/in^2.

1 kN = 10^3 N
1 MN = 10^6 N
1 GN = 10^9 N

Appendix 2 Properties of 'field' elements from the principle of variation

The derivation of the properties of finite elements to solve the field problems of section 13.2 is best illustrated by considering a specific case. Consider, for instance, the isotropic Poisson equation

$$\frac{\partial^2\phi}{\partial x^2} + \frac{\partial^2\phi}{\partial y^2} = -I \tag{2A.1}$$

which results by setting $r_x = r_y = 1$ and $I(x, y) = I$ into Eq. (13.2.3).

By applying the Euler conditions in the calculus of variations, it is easy to demonstrate that the solution of Eq. (2A.1), with its prescribed boundary conditions and within a given domain, amounts to finding the values of $\phi(x, y)$ which make the 'functional'

$$\chi = \frac{1}{2}\int_S\left(\left(\frac{\partial\phi}{\partial x}\right)^2 + \left(\frac{\partial\phi}{\partial y}\right)^2\right)\mathrm{d}x\,\mathrm{d}y - I\int_S \phi\,\mathrm{d}x\,\mathrm{d}y \tag{2A.2}$$

a minimum.

Consider the triangular element of Fig. 13.6a, for example. Assuming that $\phi(x, y)$ varies within this element according to Eq. (13.2.14), we can show by differentiation that Eq. (2A.2) becomes

$$\chi = \frac{1}{2}\int_\Delta (A_1^2 + A_2^2)\,\mathrm{d}x\,\mathrm{d}y - I\int_\Delta \phi\,\mathrm{d}x\,\mathrm{d}y \tag{2A.3}$$

in which A_1 and A_2 are the constants of Eq. (13.2.14). Now, since

$$\{A\} = \{A_0 \quad A_1 \quad A_2\}^{\mathrm{T}},$$

we can demonstrate that

$$A_1^2 + A_2^2 = \{A\}^{\mathrm{T}}\begin{bmatrix}0 & 0 & 0\\ 0 & 1 & 0\\ 0 & 0 & 1\end{bmatrix}\{A\} \tag{2A.4}$$

or, since by definition (section 9.8) $\{A\} = [c^{-1}]\{\delta\}$, we have by using the rule of transpose multiplication

$$A_1^2 + A_2^2 = \{\delta\}^{\mathrm{T}}[c^{-1}]^{\mathrm{T}}\begin{bmatrix}0 & 0 & 0\\ 0 & 1 & 0\\ 0 & 0 & 1\end{bmatrix}[c^{-1}]\{\delta\}$$

Substituting this and $\phi = \{M\}[c^{-1}]\{\delta\}$ into Eq. (2A.3) and noting that $[c^{-1}]$ and $\{\delta\}$ are independent of the current coordinates, we get

$$\chi = \tfrac{1}{2}\{\delta\}^{\mathrm{T}}[c^{-1}]^{\mathrm{T}}\begin{bmatrix}0 & 0 & 0\\ 0 & 1 & 0\\ 0 & 0 & 1\end{bmatrix}[c^{-1}]\{\delta\}\int_{\Delta}\mathrm{d}x\,\mathrm{d}y - I\{\delta\}^{\mathrm{T}}[c^{-1}]^{\mathrm{T}}\int_{\Delta}\{M\}^{\mathrm{T}}\,\mathrm{d}x\,\mathrm{d}y^{*}$$

or, since the integral $\int_{\Delta}\mathrm{d}x\,\mathrm{d}y$ taken over the entire area of the triangle is equal to its area Δ, we have

$$\chi = \tfrac{1}{2}\{\delta\}^{\mathrm{T}}[c^{-1}]^{\mathrm{T}}\begin{bmatrix}0 & 0 & 0\\ 0 & 1 & 0\\ 0 & 0 & 1\end{bmatrix}[c^{-1}]\{\delta\}\,\Delta - I\{\delta\}^{\mathrm{T}}[c^{-1}]^{\mathrm{T}}\int_{\Delta}\{M\}^{\mathrm{T}}\,\mathrm{d}x\,\mathrm{d}y$$

Then, a differentiation of this equation with respect to $\{\delta\}$ yields (see footnote of Eq. 14.5)

$$\frac{\partial\chi}{\partial\{\delta\}} = \Delta[c^{-1}]^{\mathrm{T}}\begin{bmatrix}0 & 0 & 0\\ 0 & 1 & 0\\ 0 & 0 & 1\end{bmatrix}[c^{-1}]\{\delta\} - I[c^{-1}]^{\mathrm{T}}\int_{\Delta}\{M\}^{\mathrm{T}}\,\mathrm{d}x\,\mathrm{d}y \quad (2A.5)$$

Since we have assumed $r_x = r_y = 1$, it is seen from Eq. (13.2.23) that

$$\Delta[c^{-1}]^{\mathrm{T}}\begin{bmatrix}0 & 0 & 0\\ 0 & 1 & 0\\ 0 & 0 & 1\end{bmatrix}[c^{-1}]$$

in Eq. (2A.5) is in fact the $[K_i]$ matrix of the triangular element. Furthermore, it is seen from Eq. (13.2.13) that the second term (without the negative sign) of Eq. (2A.5) is the vector $\{P_b\}$ for the triangular element in which $I(x, y) = I$.

This method is a general one in that it may be used to derive the properties of field, structural and continuum elements, provided that we are able to obtain from the calculus of variations the functional χ for the governing equation of the physical problem under consideration.

* Since $\phi(x, y)$ is scalar, we may write $\phi = \phi^{\mathrm{T}}$.

Appendix 3 Integration over a triangular area

The evaluation of integrals of the general form $\int_\Delta x^m y^n \,dx\,dy$ over a triangular area is considerably simplified when the triangular coordinates $\alpha\beta$, shown in Fig. A3.1, are introduced. It is clear from this diagram that the following

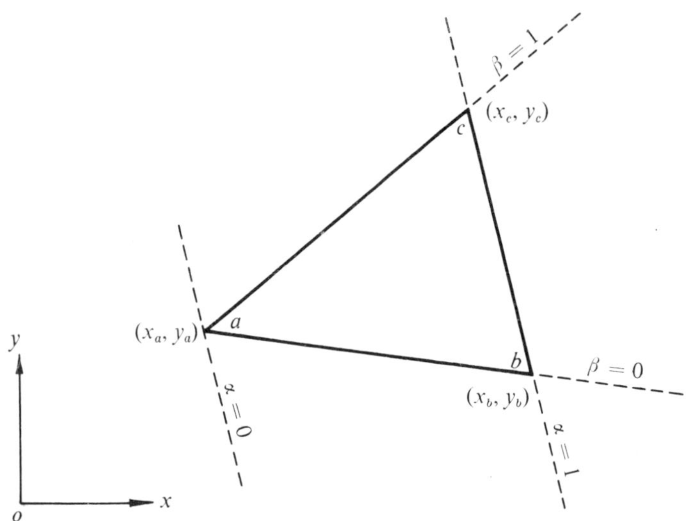

Fig. A3.1 Rectangular (yox) and triangular ($\alpha\beta$) coordinate systems.

relationships hold between the extremity coordinates in the yox and $\alpha\beta$ systems:

$$x = x_a, \quad y = y_a \quad \text{at} \quad \alpha = 0, \quad \beta = 0 \tag{A3.1a}$$

$$x = x_a, \quad y = y_a \quad \text{at} \quad \alpha = 0, \quad \beta = 1 \tag{A3.1b}$$

$$x = x_b, \quad y = y_b \quad \text{at} \quad \alpha = 1, \quad \beta = 0 \tag{A3.1c}$$

and

$$x = x_c, \quad y = y_c \quad \text{at} \quad \alpha = 1, \quad \beta = 1 \tag{A3.1d}$$

Because of these four distinct conditions, we can express each rectangular axis in terms of the triangular axes by means of four constants; in particular, let

$$x = A_0 + A_1\alpha + A_2\beta + A_3\alpha\beta \tag{A3.2}$$

By introducing the conditions of Eqs. (A3.1) into Eq. (A3.2), the constants

$A_0 \cdots A_3$ are now determined, upon which Eq. (A3.2) becomes

$$x = x_a + \alpha x_{ba} + \alpha\beta x_{cb} \tag{A3.3a}$$

in which x_{ba} and x_{cb} are the coordinate differences (section 11.1). Similarly, we can show that

$$y = y_a + \alpha y_{ba} + \alpha\beta y_{cb} \tag{A3.3b}$$

The integral $\int_\Delta x^m y^n \,\mathrm{d}x\,\mathrm{d}y$, which is in the *yox* system and carried out over the entire area of the triangle, can now be transformed to the triangular system as

$$\begin{aligned} I_{mn} &= \int_\Delta x^m y^n \,\mathrm{d}x\,\mathrm{d}y \\ &= \int_0^1\int_0^1 (x_a + \alpha x_{ba} + \alpha\beta x_{cb})^m (y_a + \alpha y_{ba} + \alpha\beta y_{cb})^n \,.\, |J| \,.\, \mathrm{d}\alpha\,\mathrm{d}\beta \end{aligned} \tag{A3.4}$$

in which the Jacobian determinant of transformation, $|J|$, is by definition

$$|J| = \begin{vmatrix} \dfrac{\partial x}{\partial \alpha} & \dfrac{\partial y}{\partial \alpha} \\ \dfrac{\partial x}{\partial \beta} & \dfrac{\partial y}{\partial \beta} \end{vmatrix} = \frac{\partial x}{\partial \alpha}\cdot\frac{\partial y}{\partial \beta} - \frac{\partial x}{\partial \beta}\cdot\frac{\partial y}{\partial \alpha} \tag{A3.5}$$

From Eqs. (A3.3), (A3.5) and section 11.1, we can demonstrate that

$$|J| = \alpha(x_{ab}y_{bc} - x_{bc}y_{ab}) = 2\,\Delta\alpha \tag{A3.6}$$

On substituting $|J|$ into Eq. (A3.4) and integrating, the following useful results are now obtained:

$$I_{00} = \int_\Delta \mathrm{d}x\,\mathrm{d}y = \Delta \tag{A3.7a}$$

$$I_{10} = \int_\Delta x\,\mathrm{d}x\,\mathrm{d}y = \frac{\Delta}{3}(x_a + x_b + x_c) \tag{A3.7b}$$

$$I_{01} = \int_\Delta y\,\mathrm{d}x\,\mathrm{d}y = \frac{\Delta}{3}(y_a + y_b + y_c) \tag{A3.7c}$$

$$I_{20} = \int_\Delta x^2\,\mathrm{d}x\,\mathrm{d}y = \frac{\Delta}{6}(x_a^2 + x_b^2 + x_c^2 + x_a x_b + x_a x_c + x_b x_c) \tag{A3.7d}$$

$$I_{02} = \int_\Delta y^2\,\mathrm{d}x\,\mathrm{d}y = \frac{\Delta}{6}(y_a^2 + y_b^2 + y_c^2 + y_a y_b + y_a y_c + y_b y_c) \tag{A3.7e}$$

$$I_{11} = \int_\Delta xy\,\mathrm{d}x\,\mathrm{d}y = \frac{\Delta}{12}(x_a y_a + x_b y_b + x_c y_c + (x_a + x_b + x_c)(y_a + y_b + y_c)) \tag{A3.7f}$$

Alternatively, if the centroid of the triangle is taken as the origin of *yox* coordinates, then

$$(x_a + x_b + x_c) = (y_a + y_b + y_c) = 0,$$

and as a result, Eqs. (A3.7) take the following simpler form in which they are more convenient to use:

$$I_{00} = \Delta \tag{A3.8a}$$

$$I_{10} = 0 \tag{A3.8b}$$

$$I_{01} = 0 \tag{A3.8c}$$

$$I_{20} = \frac{\Delta}{12}(x_a^2 + x_b^2 + x_c^2) \tag{A3.8d}$$

$$I_{02} = \frac{\Delta}{12}(y_a^2 + y_b^2 + y_c^2) \tag{A3.8e}$$

$$I_{11} = \frac{\Delta}{12}(x_a y_a + x_b y_b + x_c y_c) \tag{A3.8f}$$

Appendix 4 Some computer names used

Computer name	*Meaning*
E	Young's modulus of Elasticity.
FI	$I(x, y)$ of section 13.2.
G	Modulus of rigidity.
I	Element number.
NA,NB,NC. . . .	Nodes attached to the extremities a, b, $c \cdots$ respectively (or, $1, 2, 3 \cdots$ when the extremities are identified by digits).
NDOF	Degrees of freedom.
NEI	Total number of extremities of element I.
NEL	Total number of elements in the discretized body.
NN	Total number of nodes in the discretized body.
NP	Total number of points in element I at which the stresses are to be calculated.
NTSC	Total number of stress cards.
PR	Poisson's ratio.
RX,RY	r_x and r_y respectively of section 13.2.
T	Thickness of element I.
XA,XB,XC. . . .	x-coordinates of extremities a, b, $c \cdots$ respectively.
YA,YB,YC. . . .	y-coordinates of extremities a, b, $c \cdots$ respectively.
XP,YP etc.	Coordinates of points at which the stresses are to be calculated.